EDWARD·TELLER

BETTER A SHIELD THAN A SWORD

*Perspectives on
Defense and Technology*

THE FREE PRESS
A Division of Macmillan, Inc.
NEW YORK

Collier Macmillan Publishers
LONDON

The Free Press
A Division of Macmillan, Inc.
866 Third Avenue, New York, N. Y. 10022

Collier Macmillan Canada, Inc.

Printed in the United States of America

printing number
1   2   3   4   5   6   7   8   9   10

Library of Congress Cataloging-in-Publication Data

Teller, Edward
  Better a shield than a sword.

  Bibliography: p.
  Includes index.
    1. Teller, Edward, 1908-   . 2. Physicists—
United States—Biography. I. Title.
QC16.T37A3   1987   623.4′5119′0924   [B]   87-7584
ISBN 0-02-932461-0

Contents

Dedication

Leo Szilard
February 11, 1898–May 30, 1964

I believe that one person can change history. This book is dedicated to a man who never desired power and never attained it, but who initiated the atomic age. His memory, two decades after his death, lingers on the line between the famous and the forgotten. He was my friend Leo Szilard.

Szilard was the originator of many ideas, ranging from information theory to the sexual life of bacteria, from how to release atomic energy to a proposal that people who inform about violations of disarmament treaties ought to receive international rewards.

He was the most stimulating of all the people I have known. In a world in which conformity is almost a duty, Szilard remained a dedicated nonconformist. He had one principle, and he never violated it: He refused to say or do what was expected. For him, shocking people was a virtue; boring them was a sin. He was the only one of all my friends whom I would have liked to imitate.

It was almost impossible to embarrass Szilard. Unintentionally, I did embarrass him on one occasion. In the fall of 1945, I visited him in Santa Fe, where he shared a motel room with his girl friend of twenty years, Trudi Weiss, a physician. (He was a truly monogamous man.) Shortly afterward, I heard that they had married. When I saw him next, I congratulated him. Szilard blushed.

Like me, Szilard left Hungary because of its anti-Semitism and settled in Germany, then fled from the Nazis to the United States.

He played a unique role in American history. His ideas about atomic energy were ridiculed by Rutherford and doubted by Niels Bohr and Enrico Fermi, but accepted and acted upon by Albert Einstein and President Roosevelt.

His singular, carefully considered initiative began the American effort to develop atomic explosives. Without Szilard's intervention, work on the atomic bomb probably would have been accomplished first in the Soviet Union, and freedom might have been long since extinguished. Yet, after World War II, Szilard came to believe that the tension between the United States and the Soviet Union was due to various misunderstandings rather than to fundamental differences.

There we disagreed completely. Still our friendship, which was strong but strangely unemotional, survived. Some of my most pleasant memories of Szilard are connected with our debates, many of them private, some of them public. Szilard would have liked my evasion far less than my open opposition to his ideas. He was a great liberal in the oldest and best sense of the word. He was tolerant of other people's ideas as long as his own ideas could be expressed. The great advantage of such an arrangement is that it puts agreements and differences into proper perspective.

Szilard and I easily agreed that we wanted peace. We readily agreed that peace was endangered and that the outlook was dark. We even agreed that ultimately we must cooperate with everyone on the globe, including the Soviets. That helped to bring out more clearly our important but limited disagreement, which concerned tactics: Shall we first become strong so that we can become pleasant later, or could we immediately be pleasant and sidestep competition with the Soviets?

One incident illustrates why Szilard never became truly famous or influential. He participated in the Pugwash Conferences, begun by Cyrus Eaton for the purpose of saving the world by combining the wisdom of scientists from all countries. At the first conference, Szilard proposed that the faults of both the American and Soviet systems be clearly outlined. He suggested that the task be taken up by the true experts in each field: The "horrors" of American capitalism should be exposed by Americans, and the "difficulties" in the Soviet system should be described by the Soviets. His suggestion was met by silence and then was tabled. Yet what could have provided a more refreshing beginning for a meaningful debate?

Szilard had no trace of personal ambition, which he considered a childish folly. He was, however, fond of giving advice, and his advice sounded like a command. He acquired a nickname among his friends—The General. That goes far toward explaining his unpopularity among those who had power or influence. He criticized others too often for his own good, and not tactfully enough for the good of those whom he criticized. But unlike Socrates, Szilard was not condemned to drink the poisoned cup, which suggests that freedom of speech is respected more in the United States than it was in ancient Athens.

In a democratic society, a man like Leo Szilard can have an influence no matter what his opinions. In a totalitarian society, he could end up in a psychiatric institution. The opportunity to make a contribution exists for each of us, so long as we maintain openness in our society.

In historical terms, mankind acquired such openness very recently. Freedom is a miracle. We shall not retain it for many more years unless we realize that it has to be defended from day to day. By what we accomplish or fail to do, each of us has an influence, just as Szilard did, on the future of mankind.

As I write this, I have just returned from another trip to Europe. I have been arguing for attempting to make weapons of mass destruction obsolete. Leo Szilard would not have been neutral on this question. I do not know which side he would have taken, but I wish I could hear his unpredictable arguments, which never bored anyone.

EDWARD TELLER
Hoover Institution
Stanford, California

Acknowledgments

Ideas and beliefs have a history. In the case of many of the ideas in this volume, years passed in conversation and thought before I lectured or published on the topic. Quite a few statements in this book have not been put into concrete form on any earlier occasion. Putting these ideas into print has helped their consistency and substance.

I am indebted to many friends for the results. I cannot mention all of them, but those from whom I obtained insights I otherwise would have missed deserve special mention.

Lowell Wood convinced me that true originality continues to exist. More than one of the valid proposals in the Strategic Defense Initiative are his. His record is equal to that of the best I knew in the Manhattan project.

George A. Keyworth was the science adviser of President Reagan in the first stages of the SDI. I am indebted to Jay for an understanding of the real novelty of the Strategic Defense Initiative—that defense must be international.

Arthur Kantrowitz has sustained my faith in innovative technology. His accomplishments stretch from hydrodynamics and the reentry of vehicles from space, to the invention of a device that helps the human heart pump blood and to his carbon dioxide laser, which is used throughout the world. I am indebted to him specifically for discussion of technology under the Ming dynasty and for suggestions about how to decrease secrecy in the United States.

During most of the long period over which these thoughts have gestated, Hans Mark has been my close friend—from his early days as a student of atomic physics, when he built a nuclear reactor in Berkeley, through his leadership of the NASA laboratory at Ames, to his contributions as Secretary of the Air Force and as deputy administrator of NASA. Now, as Chancellor of the University of Texas, he will have an important impact on how science and technology will appear to my grandchildren. I thank Hans for a sense of balance at a time when there seems to be more controversy than understanding.

As a professor and public speaker, I feel that I handle words adequately, although not perfectly. But when thoughts are to be written, a more careful use of words is imperative. I thank Judy Shoolery for helping me to find the words and particularly for her patience, which was necessary, in steering me away from major collisions with grammar. I have learned a lot from her about the long road that leads from mere ideas to black print.

A Personal Note About War

It was Saturday morning, July 14, 1928. The ambulance had arrived to take me to the hospital in Munich. At the age of twenty, for lack of common sense, I had been involved in a streetcar accident. I was to go through life without a right foot.

The pain was acute and unrelenting. I remembered, from my childhood, stories of the terrible world war that had ended ten years before, and I felt lucky by comparison. I knew there soon would be help. The war-injured seldom have help so quickly.

At no time during my long life have I been able to forget what war really brings. I saw the violence and hatred of civil war. I knew hunger and separation from family and friends. I learned a lot about war before I reached twelve, and since then the fear of war is a reality that never leaves me.

On that Saturday morning in 1928, kind and competent Dr. von Lossow eased much of my suffering with the amputation. Before I was fully recovered, he disappeared without saying goodby. Much later I learned that he was the brother of General von Lossow, who had arrested Adolf Hitler five years earlier during the abortive Munich Putsch. The Nazi party was having a resurgence during the summer of my recuperation. Dr. von Lossow, knowing that Hitler hated his brother, left his homeland as quietly and quickly as he could.

But before the Nazi terror descended on Germany, I had a chance to learn about the world of atoms. I knew next to nothing about politics, but I found out what I could about theoretical physics. I also read some poetry. One of my favorite poets was Erich Kästner. His poems, in colloquial German, frequently took unexpected and frightening turns. I have made attempts at translating one that was published the year following my accident:

I have no dreams—that I can freely confess.
When I sleep, I am quiet, unconscious, and dumb.
But one dream I remember, nevertheless.
It was of the war that will come.

Men rose from trenches all over the land.
(Volunteers, I was assured by a voice.)
Their rifles ready for the dreadful command,
But whom to kill was everyone's choice.

They walked like zombies. None made a sound.
Then came a cry as from someone in pain.
All men turned their rifles around.
And, by himself, each one was slain.

At once a million lives were undone.
When I sleep, I never must dream.
But I wish that I knew who was the one,
The one who did scream.[1]

During the beginning of my life in science, I saw all around me the developments that led to disaster and to World War II. The fears and hopes of the second half of our century are a strangely transmuted repetition of the fears and hopes of the first half.

Two years after Hitler took power, I came to America. I wanted to escape politics and continue to study the tiny entities of which matter is built. But the collapse of peace was not limited to Europe. I became a part of the atomic age in all its practical phases, from constructing explosives to ensuring the safety of reactors.

The debate about disarmament, defense, and peace is today more persistent and pervasive than ever before. The problem itself is ancient. As for myself, it has preoccupied my conscious existence.

I

THE STRATEGIC DEFENSE INITIATIVE

1

Three Controversies

Should we have built the atomic bomb? Should the bomb used at Hiroshima have been followed by the development of another one, a thousand times more powerful? Should we establish a defense against missiles carrying such bombs?

Of those three controversies, the first arose not before the weapon was developed but many years after the bomb had been used. The second controversy was ended by a Presidential decision, but its repercussions have not subsided. The controversy over strategic defense is raging today. In all three debates, many of the same people are participants. That is to be expected. The variety of the arguments is more surprising.

THE ATOMIC BOMB

Intensive work on the atomic bomb began during the darkest hours of World War II. Not only did Vice President Truman know nothing of the project, but the nature of the work was kept secret even from many of the people working on the Los Alamos mesa.

One evening during a trip down from the laboratory, the Fifty-Minute WAC (a woman soldier who could drive any distance in fifty minutes) glumly commented: "My girl friend will come home from Africa and tell our acquaintances, 'I drove a munitions truck to the crucial battle of El Alamein.' I will say, 'I stayed in the States and drove Mr. Teller [even the fact that I had a doctorate in physics was classified] to Santa Fe.'" Not until after Hiroshima was announced did I see the Fifty-Minute WAC smile.

Forty years later, not many people smile. Many of the scientists who planned the first atomic bombs made up their minds never to work on weapons again. Before Hiroshima, one of those scientists

3

recommended that the United States wait until a dozen atomic bombs were available and then drop them all on the same day; that would surely win the war. Today he opposes Strategic Defense.

In retrospect, it is clear that no atomic bomb should have been dropped before its incredible power had been demonstrated to the Japanese. Hiroshima did not change a single vote in the Japanese War Cabinet, which was responsible for decisions about the war. It did affect Emperor Hirohito, who broke with tradition and the Japanese constitution and appealed directly to the people for peace. A demonstration of the bomb and the power of science, 30,000 feet above Tokyo Bay, would not have hurt anyone. It would have been seen by Hirohito, would have ended the war, and would have left behind a world more reasonable and secure.

We developed the bomb in fear that the Nazis might have it first. Actually, the Germans under the leadership of Werner Heisenberg barely tried. Soviet work on the project was curtailed by wartime shortages. But had not a democracy been the first to succeed, it is a foregone conclusion that a totalitarian state would have succeeded not many years later. Critics of the atomic bomb development fail to mention that our inaction would have set the stage for world dictatorship.

The United States as sole possessors of the first atomic weapons did not use their immense new power to create a world empire. Restraint was exercised because the first development took place in a democracy. I do not claim that the United States makes no mistakes, but only that in our country discussion and reason have a chance.

Increasing knowledge and progress, including progress in the military field, are unavoidable. Los Alamos did not change the world by developing the atomic bomb. It did change world history by placing the first bomb in the hands of those who value peace more than power.

THE HYDROGEN BOMB

Why develop the hydrogen bomb? Wasn't the atomic bomb destructive enough?

Obviously, weapons have been too destructive since a man first used a rock to kill another man. A more pertinent question concerns ignorance. Can lack of knowledge ever contribute to stability or safety? The hydrogen bomb controversy marked the first time that a large group of scientists argued for remaining ignorant of technical possibilities.

But if the United States ceased to develop destructive weapons, would others follow suit? In late October 1949, that hope was formally raised by the General Advisory Committee (GAC) of the Atomic Energy Committee. According to Andrei Sakharov, the physicist instrumental in the development of the Soviet hydrogen bomb, the Soviet effort had been under way for at least a year and a half when the American hope of delaying development was voiced.[1]

The public debate on the hydrogen bomb ended after a few weeks with President Truman's decision to proceed with development. At the beginning of the debate, two great scientists, Enrico Fermi and I. I. Rabi, made a statement that had a lasting effect on public opinion. They said: "The fact that no limit exists to the destructiveness of this weapon makes its very existence and the knowledge of its construction a danger to humanity as a whole."[2]

The much greater explosive yield of a thermonuclear weapon proved to be an alarming reality—the thousandfold increase in explosive power provided by the atomic bomb was followed by a further thousandfold increase. However, contrary to expectation, destructiveness does have limits. The limits are both technical and deliberate.

A very big explosive blows the atmosphere within a diameter of 10 miles into space. With a further increase in the size of the explosive, practically the same mass of air will be blown into space at a higher velocity. The lateral effects along the ground expand to an exceedingly small extent. Thus, further escalation in the power of the explosive is ineffectual and useless.

Actually, the megatonnage in the U.S. stockpile peaked in the mid-1960s and has decreased steadily from that time to one-quarter of its maximum value as the accompanying graph shows. Current U.S. weapons do not include any of multimegaton size. The Soviet nuclear stockpile has increased in megatonnage and exceeds the U.S. explosive power severalfold at this time.

An additional argument against developing the hydrogen bomb was the critics' claim that fusion, unlike fission, had no potential peacetime applications. Other arguments were that clusters of fission bombs were militarily more effective; that the hydrogen bomb was unnecessary because plans were already under way to build an atomic bomb with a 500,000-ton equivalent TNT yield;[3] that the fusion

Total Yields of U.S. and USSR Stockpiles, 1965–1985

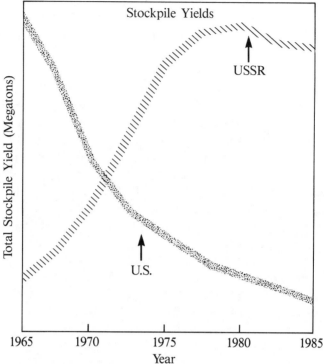

Source: U.S. Department of Defense, 1986

weapon might be impossible to develop; and that even if possible, it would be unwieldy and expensive.

The thermonuclear explosive produces less radioactivity. It can therefore be used in construction and mining work, as the Soviets have demonstrated. Contrary to predictions, the thermonuclear weapon turned out to be less expensive and much smaller in weight and size. That meant rockets could easily be used for their delivery, a fact that caused the United States to make a large-scale effort to catch up with Soviet rocket technology.

One final argument against development was that if the Soviets did produce a thermonuclear weapon, the United States could close the gap quickly. How we would gauge Russian progress or gather support for matching a secret Soviet effort was not specified.

No one can be happy about the balance of terror that followed the development of thermonuclear weapons, but a Soviet monopoly on intercontinental missiles carrying thermonuclear weapons would have been worse.

THE STRATEGIC DEFENSE INITIATIVE

The Soviet Union has emphasized real defense in all its various forms since the early 1950s. In view of the suffering incurred during World War II, that is hardly surprising. Considering the thorough Soviet emphasis on military affairs, one would expect its air defenses to be superior (as they are), its civil defense to be as complete as possible (which it is), and, given its early entry into rocketry, that the Soviet Union would lead in ballistic missile defense (which has been deployed around Moscow and twice upgraded).

Nevertheless, the Soviet reaction to Reagan's proposal for American defense was prompt and negative. To understand the SDI controversy, which in many respects is more ferocious than was the hydrogen bomb debate, its beginning must be examined.

On March 23, 1983, the President announced "a long-term research and development program to begin to achieve our ultimate goal of eliminating the threat posed by strategic nuclear missiles."[4] He did not mention stars, space, or satellites.

The immediate reports in the American press quoted Teddy Kennedy rather than Ronald Reagan. The headlines cried: "Star Wars!" The central question of how best to deter war in the face of new technical developments was lost in a discussion of space battle stations. The President had specified the essential issue, but the media chose to pursue an unmentioned, peripheral, sensational question.

On March 26, General Secretary Andropov criticized the Strategic Defense Initiative as an effort "to militarize space." That response fitted only the press reports, not the President's speech. Two weeks later, the entire membership of the Soviet Academy of Sciences signed a statement opposing the "destabilizing" Star Wars program.[5] That cue was picked up by many American academics and, throughout the free democracies, by all parties to the left of center. The actual Strategic Defense Initiative has been hidden by misunderstanding and media misrepresentation. What is SDI, and why do we need it?

For many years, the only defense that received serious attention in the United States was the destruction of incoming rockets at a time shortly before they reached their targets. That is the type of defense established around Moscow. Work on such a system started in the United States in the 1960s. Actual deployment was begun in the 1970s and then abandoned. SDI continues the old attempt but emphasizes technologies developed in the last two decades.

A new emphasis on lasers and other directed-energy weapons has proved important. The most common lasers are visible light beams of sharp definition and enormous intensity. Other types range from the long-wavelength microwave laser to the exceedingly short-wavelength, novel, and remarkable x-ray laser.

What is common to the new defensive weapons is that they are sharply directed against individual objects, in particular against incoming missiles. The SDI research program includes no weapons that could produce destruction in any way approximating the massive effects of current high-yield nuclear weapons. Weapons of mass destruction are indiscriminate; directed-energy weapons are highly discriminatory.

Defensive weapons may be deployed in appropriate orbits. They also may be kept on the ground. Ground-based defensive weapons can provide defense against missiles launched from distances of less than 500 miles, for example, from offshore submarines. The range of the defensive weapons can be extended by mirrors popped up into an appropriate position as soon as a warning of attack is received, or the defensive weapon itself may be popped up with sufficient speed. Predeployed weapons in space are by no means necessary for defense.

Another novel emphasis has been consideration of means for attacking missiles shortly after they are launched, while they are being accelerated. In this early stage, several missiles and many decoys are apt to be carried by a single rocket, so destroying the rocket then is particularly advantageous. Soviet progress in those fields—particularly in research pertinent to directed-energy weapons—makes the urgent need for SDI obvious.

Any person with a humane point of view should be opposed to aggression, but why be opposed to defense? The most popular argument is that defense against nuclear missiles is useless unless it is 100 percent effective—the damage would be enormous if only a few of the rockets penetrated the defense. The argument is correct, but it is an argument not against defense but against war. No matter what weapons are involved, any full-scale war will be devastating, as wars in the recent and remote past have been.

The main purpose of defense is to deter aggression by making its outcome dubious. In that sense, defense alone may not suffice, but defense may well make an important contribution to deterrence. Furthermore, should deterrence fail in spite of every effort, the existence of defense at least will save lives.

A further argument against defense is that it is destabilizing. If that is so, the protests should be directed primarily against the Soviet Union. Soviet leaders have deployed both terminal defenses around Moscow and laser defense at the Sary Shagan weapons test site near Lake Balkhash. The argument that defense is destabilizing gained widespread acceptance only after Reagan's proposal was criticized by the Soviet Academy of Sciences. It has not been applied to the past or present intensive Soviet defense effort.

What about the expense? A budget of $26 billion has been proposed for SDI. The qualification that this expenditure would support five years research is rarely mentioned. In the end it will amount to less than 2 percent of our total military appropriations.

Opponents of defense remind us that successful defense will not be the ultimate answer, any more than previous ultimate weapons, such as the atomic and hydrogen bombs, have been. In particular, they claim that more defense will stimulate additional aggressive deployment. Thus, work on armaments would proceed indefinitely.

In principle, that argument is correct. In practice, our immediate aim is to overcome the present predominance of weapons of mass destruction. SDI is restricted to research. Attempts to deploy defense will be made only if there is a chance that defense will be less expensive than the compensatory deployment of aggressive weapons or the means to destroy the defenses. There can be no claim, even if we find a solution, that it will be permanent.

However, defense has the advantage that it can provide the basis for a decrease in tensions. In the end, peace must be much more than the absence of war or the absence of the technical means to carry out war. Peace must be founded on peaceful intentions, and, therefore, on cooperative actions. The real question is whether it makes sense to aim for such a solution now or whether it is better to make a start by strengthening the possibility of peace and by decreasing the fear that is rightly associated with weapons of mass destruction. Actually, the President's proposal—to share the results of our defensive efforts with everyone, including the Soviet Union—is the most realistic political step yet proposed. Eventually it could lead to a truly peaceful atmosphere.

For three decades, the United States has tried to make progress toward peace by seeking political solutions. It is impossible to assert that the nation is closer to peace now than in the 1950s. Whether the worsening of the situation is ascribed to the futility of the arms race or to the futility of disarmament negotiations depends on one's point of view.

THEN AND NOW

There are significant differences between the hydrogen bomb debate and the current SDI debate. Today, the scientists opposed to research and development have been out of touch with technical progress for a far longer period of time. Some of them have never participated. The young people who are the heaviest contributors of the new ideas, because their work is classified, are not properly recognized within the scientific community, much less among the general public.

Another difference is that the SDI debate is being conducted in the international rather than the national arena. General Secretary Gorbachev is presenting the position of the Soviet Union in a persistent manner. It is obvious that throughout the democracies, the left wing favors Gorbachev and the right wing favors defense. What is more peculiar is that the opponents of defense now point to the advantages of mutual assured destruction. Three years ago, those same people were protesting against retaliatory power.

The greatest difference, however, is more ominous. In 1949, the United States had the strongest military arsenal in the world. Had we made a mistake then and recognized the peril in the early 1950s, we might possibly have caught up again. In 1985, the Soviet Union is stronger in every quantitative sense. If we make a mistake today, we have no margin to absorb the consequences.

In 1949, I advocated work on the hydrogen bomb, a weapon of attack. I am now arguing for the development of the means to defend against those weapons. The change is due to new understandings, which have arisen from increased technical knowledge. Now as then, my argument is for knowledge and against ignorance. Now as then, I offer no detailed proposals as to how the knowledge, once acquired, should be used. Scientists have the responsibility to make knowledge available and to explain its possible applications. The decision as to which uses should be adopted, now as then, should belong to the entire community. That is indeed the main principle on which a democratic society rests.

Those are some of the parallels and peculiarities of three controversies. Comparing and contrasting them is an exercise in remembering the past and clarifying our current perspectives.

2

Better a Shield than a Sword

Soviet Defense

I think that a defensive system that prevents attack is not a cause of the arms race. . . . Perhaps an antimissile system is more expensive than an offensive system, but its purpose is not to kill people, but to save human lives.[1]

I agree wholeheartedly. As long as there are two opposing camps, the world is in less danger if they are armed with shields rather than with swords.

The statement above was made in London by Aleksei N. Kosygin, premier of the Soviet Union, on February 9, 1967. This part of Soviet policy has never changed.

In 1985 the book *History Teaches Vigilance* was approved for publication and praised by the Soviet press. The author, Soviet Marshall N. V. Ogarkov, reiterates the opinion that defense, as well as offense, is of crucial importance and emphasizes that "the appearance of new means of attack inevitably leads to the creation of corresponding means to counter them."[2]

Today a widespread controversy is raging about whether the threat of retaliation is a unique and necessary deterrent of war. Discussion of that subject should include a consideration of the development of Soviet and American defensive weapons prior to the Strategic Defense Initiative.

World War II, according to Soviet records, took the lives of 20,000,000 people in the Soviet Union. Yet, during the war, I. V. Kurchatov, a great Soviet physicist, started work on atomic explosives with a few dozen collaborators and managed to understand the problems and establish elementary tools. By 1945, the work of his group had advanced to a stage comparable to that in the United States in 1942.

The actual nuclear explosion in Japan came as a shock. Kurchatov's biographer wrote:

> Hopes for a lasting peace and the triumph of reason had not had time to be confirmed when . . . explosions resounded in Hiroshima and Nagasaki that were echoed in diplomatic encounters. The American administration had conceived the idea of dictating its will to peoples who had defended freedom in the bloodiest war in the history of mankind. The most unrestrained atomic blackmail began and with it the "cold" war waged by the United States of America against its ally of yesterday—the Soviet Union.[3]

In the Western view, "the idea of dictating . . . to peoples who had defended freedom" consisted of protesting against Soviet expansion into Eastern Europe and northern Iran. The Soviets actually withdrew from northern Iran.

After a delay of almost a year, the Truman Administration proposed the Baruch Plan to internationalize both peaceful and military uses of atomic energy with guarantees that no secret work would proceed. Among all the treaties ever proposed, that was the only one with sufficient guarantees to give it a chance of becoming effective. It met with prompt rejection by the Soviets under Stalin.

The effort to create nuclear weapons became the Soviets' highest priority goal. In 1949, sooner than anyone had predicted, they had an atomic bomb. This fact in itself should have been sufficient evidence that, as has been true of Russia since the reign of Peter the Great, in military technology the Soviet Union is a fully developed nation.

Considering their history, the Soviets' emphasis on a strong defense can be understood. Because the new bombs were delivered by airplane, the next step was the development of air defenses. Deployment of fighter planes began soon after the Soviet atomic bomb was tested. During the late 1950s, the Soviet National Air Defense became an independent military service; by 1959 it was surpassed in importance only by the Strategic Rocket Forces and the Ground Forces.

Today, in 1987, the deployed conventional air defense protecting the Soviet Union includes thousands of fighter-interceptor aircraft armed with air-to-air missiles and more thousands of surface-to-air missiles and launchers. Those defensive aircraft seem to have caused little concern in the United States, because the technical quality of Soviet equipment is assumed to be inferior.

In 1976, a high-performance fighter (the MiG-25) was flown to Japan by a Soviet defector. Analysts who disassembled the plane commented on its low technological level. Rivets on the air frame were not well smoothed; the aerodynamic surfaces were not optimally shaped. It attained its great speed—an impressive two and one-half times the speed of sound—simply by means of a high thrust-to-weight ratio. Furthermore, as the American press happily reported, the plane had voluminous vacuum tube electronics rather than modern miniaturized chips.

A few weeks later, unnoticed by most of the public, the evaluation of Soviet technology had to be reversed. Deep within the plane, in shielded positions, the Soviets had used transistors. Old-fashioned tubes have the advantage of being resistant to nuclear radiation. Modern electronic equipment is too sensitive for use in nuclear combat. More than a decade ago, the Soviets had already developed superbly effective aircraft appropriate for nuclear conflict.

Today, the Soviets have more than a thousand interceptor aircraft assigned to territorial strategic defense, and almost three thousand more assigned to the Air Force could play a similar role. Some Soviet defense aircraft have "look down/shoot down" capabilities important in defense against low-flying aircraft and cruise missiles, which are hard to detect from the ground.

Surface-to-air missiles (SAM) may be even more important for defense than fighters. Every few years since the early 1970s, the Soviets have produced a new generation of those rockets. The Soviet SA-6 missile appeared in the hands of the Egyptians and Syrians during the 1973 Yom Kippur War. That missile combined command guidance, radar-reflection homing, and infrared homing. None of the electronics was novel or particularly advanced, but it provides another example of modest technologies used in an innovative, efficient combination. The missile prevented the effective use of the Israeli air force during the nearly decisive first few days of the conflict.

Soviet military planners today have a wide variety of such surface-to-air defense rockets. The SA-10, clearly effective against bombers, also may be useful against ballistic missiles, although whether they were developed specifically for the latter purpose is debatable. A further system, the SA-12, with widespread deployment expected by 1990, is even more likely to have capabilities against ballistic missiles. The system is mobile and fires very high-acceleration interceptor rockets. A fast-accelerating rocket, armed with a very small nuclear device (possibly equivalent to 10 tons of TNT), would provide effective defense against an intercontinental missile. The fact that the whole defense is mobile makes it possible to change the configuration of deployment. Today the Soviets have deployed 12,000 SAM launchers and 10,000 radars supporting the SA-12 type of air defense.

About 1956, Soviet research began on ballistic missile defense and the necessary radar. Although the rocket age did not begin until a year later, when the Soviets launched the first space capsule, Soviet military leaders were quick to see that the rocket as well as the airplane could provide a delivery system for the new nuclear weapons. In 1961 and 1962, even before many rockets were deployed, the Soviets conducted nuclear tests in the atmosphere pertinent to missile defense. Based on those experiments, in 1965 the Soviets began deploying their first anti-ballistic missile (ABM) defense system around Moscow. In 1967 the deployment of the supporting radars began.[4]

The original system was a single-layer defense that used interceptor rockets armed with small nuclear explosives. Those could destroy incoming rockets if detonated in the general vicinity of the warheads as they reentered the atmosphere. That is an advantage over interceptors that must score a direct hit.

In the fall of 1986 Arkady N. Shevchenko, the highest-ranking Soviet diplomat ever to defect, gave a talk at Lawrence Livermore National Laboratory. Shevchenko told us that the official Soviet estimates at the end of the 1970s assumed that the defenses then deployed around Moscow could stop more than half of the attacking rockets. After scheduled improvements, the system was expected to be able to destroy more than ninety percent of the incoming rockets aimed at Moscow. The rocket system protecting Moscow has now been upgraded. American retaliatory missiles remain unchanged.

The Soviet Union has possessed a functional missile defense system for more than a decade, which has given them a chance to work out all the engineering problems. They almost certainly have many people trained in deploying and operating the terminal defense system. The new system installed around Moscow uses components that can be easily duplicated and rapidly deployed.

On the other hand, it takes a long time to deploy the expensive, extensive radar equipment crucial to ABM defense. With the deployment of the Krasnoyarsk radar station, the last link in the Soviet radar net will be completed. Defense of many cities in the Soviet Union could be realized in the near future. While Soviet leaders continue to protest against American research on defense, their expenditures on deployment of defense radar proceeds unabated. Six other radar stations, able to track incoming missiles with greater accuracy, are under construction to replace older versions. The data transmission facilities linking those stations to other key defense commands are being modernized. The Soviets claim that the purpose of those radars is air defense. Indeed, such radar may serve a multiplicity of purposes.

Ten years ago in the United States, only terminal ABM defense seemed feasible. Its advantage is that heavy, expensive rockets can be easily differentiated from decoys after they reenter the atmosphere. But at that late stage only a short time is available for defense. Intercontinental missiles may be attacked in a more leisurely manner during the half-hour they spend in space, but they are then hard to distinguish from decoys. Actually, they would be attacked most effectively during the boost phase, shortly after they have been fired.

In the boost stage, one attack rocket is apt to carry several nuclear explosives, which may reenter the atmosphere at widely separated locations. Each rocket also may carry a hundred or more decoys as well as chaff—small objects that obstruct radar observation. The mechanism accelerating the rocket is vulnerable, and even small damage is likely to lead to an explosion and the destruction of the rocket and its contents.

The Soviets have tackled the problems of boost phase and midcourse interception for almost twenty years. Since 1968, they have worked on directed-energy weapons that fire with the velocity of light and need only about one-hundredth of a second to reach a target a couple of thousand miles away. The best-known form of a directed-energy weapon is an intense light beam, called a laser, which can be aimed with an accuracy better than one part in a million. Sufficiently intense lasers can destroy any object they hit. An observation satellite is an obvious target.

In 1987, the Soviets have several major research and development facilities for laser work, employing more than 10,000 scientists and engineers. A comparable laser research program in the United States would, by itself, cost $1 billion a year. The Soviets have deployed a powerful laser at Sary Shagan, a military test site that has long been the center for ABM research.

Soviet research has contributed many original scientific ideas that upon examination have proved to have potential relevance to defense. For example, lightweight equipment to accelerate atomic particles to nearly the speed of light, and techniques for accelerating and aiming electrons through a narrow channel previously established in a near vacuum by an intensive laser beam are both ideas derived from Soviet research reports. In 1978, a paper describing a 1.2-meter segmented mirror was published. Such mirrors are needed in connection with the use of powerful lasers. The Soviets also have published preliminary research on the novel and exciting x-ray laser. In the late 1970s, the publications abruptly stopped.

Such innovative research has not been confined to defensive weapons. A very different development of nuclear weapons was initiated in 1966 by Andrei Sakharov, the greatest of the Russian military scientists. A team led by Sakharov stated that it had produced a magnetic field of 25 million Gauss. Such a field is so rich in energy that a cubic inch contains about thirty times the energy of a good high explosive of the same volume. It is probable that the high magnetic field was produced by a nuclear explosion.

Transformation of the energy of a nuclear explosion into electric or magnetic energy may become important. The Soviets have shown a systematic interest in the high-intensity radiation that accompanies nuclear explosions. In a U.S. high-altitude test in 1962, we were surprised by an observation of such an electromagnetic pulse (EMP). We found the explanation in the open Soviet literature. EMP acts like lightning and can damage communications and power transmission in a wide area. Radiation of shorter wavelength can destroy small, sensitive electronic devices, which are extremely widespread in the United States.

If the Soviets have found a way to transform a considerable fraction of the energy produced by a nuclear explosion into a downward-directed EMP, the resulting damage would paralyze communications and transportation systems over a wide territory. The United States relies increasingly on electronics. Defense against such a weapon would be difficult and expensive. In the United States, work on, and even understanding of, this phenomenon is incomplete.

In addition to all active measures of preparation, the Soviet Union has given high priority to the passive defense of its officials and of its people. Soviet civil defense includes appropriate hardened shelters for 175,000 party and military personnel, storage of key materials underground, relocation plans for industry, and evacuation plans for civilians. In 1984, the Soviet Union spent $13.00 per capita on passive defense; the United States, $0.78. No one should criticize defensive measures, whether they are pursued in Switzerland, the Soviet Union, or the United States.

American Defense

Prior to 1983, development of defense in the Soviet Union and the United States took basically different directions: the one successful, the other unsuccessful; the one logical, the other haphazard; the one justified, the other irresponsible. In all those respects, the Soviet Union was right, the United States wrong. The historical background is easily described but difficult to accept.

Between the discovery of fission and the first atomic bomb, less than seven years elapsed. Today, no one dares to propose a comparably rapid novel program. In 1945 a belief in American technical superiority was justified. Today it is not.

In the early 1950s, the United States, like the Soviet Union, began to develop defenses against aircraft. That decision was made on a bipartisan basis. Doves and hawks agreed that defense was needed to ensure peace and make disaster less likely.

In 1957 the Soviets launched the first space capsule, thereby accomplishing the unthinkable feat of taking the lead in a technological development. The initial response to the surprise of Sputnik was to speed up the U.S. rocket effort. The first production, the "Sitznik," failed to get off the pad. But eventually American technology led by Si Ramo and Air Force efforts organized by Colonel Benny Shriever produced miracles.[5] President Kennedy, over the objection of many scientists, initiated the program to land a man on the moon. It was an open program, where neither successes nor failures could be concealed. The manned space program became a great American success.

An unhappy consequence was a further change in the military situation. Airplanes could cross the ocean in hours; missiles needed no more than thirty minutes. In addition, the development of the hydrogen bomb had made nuclear explosives not only more powerful but also easier to deliver by rockets. The net effect seemed to give attack an advantage over defense. Airplane defense appeared insufficient and was abandoned. By the late 1950s, the prospects for providing effective comprehensive defense were considered hopeless. At first I agreed with that view. In 1960 I became convinced through detailed discussions with Air Force General E. E. Partridge that rocket defense must be attempted.

About four years after the Soviet effort began, the United States started work on high-acceleration missiles able to intercept incoming rockets. If the interceptor, called Sprint, could come near the attack missile, a nuclear explosion of appropriate size could incapacitate the explosion mechanism in the enemy warhead. The rockets would approach faster than a rifle bullet. Still, it is not justified to say that rocket defense is as difficult as hitting a bullet with a bullet.

At first glance, it seems absurd to plan defensive nuclear explosions above one's own territory as the Soviets have. I know, not only from theory but also from personal experience, that it is not absurd. In the early 1950s, I once sat on the glass bottom of an observation plane over southern Nevada at dawn, watching a fixed spot on the ground where a nuclear explosive with a yield equivalent to a few thousand tons of TNT was to be set off. It went off on schedule. Apart from a short flash, it looked like a small amount of mud stirred up by a pebble at the bottom of a deep pond of clear water. Had I not watched for it, I would have missed it, even though I was sitting 5 miles straight above it. The shock wave was due to arrive about half a minute after the explosion. It came as expected, but, even anticipating it, I could barely feel it over the normal bumpiness of the flight. The subsequent mushroom cloud was monstrous. It grew to a height a little above our plane. Had we not avoided it, we would have been enveloped.

If a similar explosive were detonated 5 miles above the ground, the mushroom cloud, composed of dust and debris stirred up by the explosion and carried aloft, would not be produced. At 25,000 feet, such an explosion might alter or form clouds of the sort seen in the wake of a jet airplane. A shock wave (or sound) becomes stronger as it travels upward and weaker as it travels downward. That is a law of nature as sure as the statement that a stone will not fall upward. A small nuclear explosion, such as would be used in defense, could scarcely be heard 5 miles below.

One serious difficulty of ABM defense is that it requires extensive, expensive, vulnerable radar installations. If the radar is put out of action, the defender is helpless. Shielding radar against both the shock wave and the EMP that the explosion of a large warhead would produce is a formidable problem. It was the Achilles' heel of the defense planned in the 1960s.

Research and initial development of terminal defense received considerable attention in the 1960s. An effort about one-quarter as great as the SDI was applied exclusively to providing terminal defense. That defense was based on an appropriate nuclear explosive, which would render the attacking rocket harmless after the rocket had reentered the atmosphere and the accompanying decoys had been eliminated. In 1969, deployment of such a defense was contemplated. My friend Henry Kissinger, who was then the National Security Adviser to President Nixon, telephoned: "Edward, you argued for the anti-ballistic missile. We are supporting it, but there will be objections. You must go on the road and defend it." So I did.

In June of that year, I was invited to speak at a conference of the media of Montana at Glacier National Park. I arrived on a beautiful, sunny morning and went for a walk with some other members of the conference. That afternoon, my talk was introduced by one of my companions. He mentioned our little excursion: "A hundred yards from the lodge, Dr. Teller picked up a large stick. I asked him what it was for. He replied, 'For protection against grizzly bears.' I protested, 'That stick isn't effective enough!' He answered, 'I know, but I hope the grizzly bears don't.' Edward Teller will now talk about ABM defense." That was, at the time, a justifiable introduction.

The 1969 ABM debate ended with approval by the margin of a single vote in the Senate. Many continued to prefer Mutual Assured Destruction, appropriately abbreviated MAD. The logical mind of Secretary of Defense Robert Strange McNamara conceived, named, and formulated that plan in the early 1960s. That policy claimed the only security available was to make nuclear war truly unthinkable by concentrating on retaliation and abandoning, even forbidding, defense.

Following the Senate vote, actual deployment of terminal defense to guard our retaliatory missiles began in North Dakota. But work was scarcely started when the American ABM effort was used as a bargaining chip. In 1972, missile defense was limited by a treaty signed in Moscow. There would be one defensive site permitted in the United States and one in the Soviet Union. The Soviets defended the people of Moscow. We defended the missiles in North Dakota. After a short period our defense effort was stopped. The Soviet effort continued.

In that manner, MAD became a unilateral American policy. Apart from a limited research effort on the old type of terminal defense, interest in methods of stopping a rocket attack or any other attack by weapons of mass destruction was dead. Defense of the American people by active or passive methods became a nonissue.

THE STRATEGIC DEFENSE INITIATIVE

When Ronald Reagan became President of the United States, one of his early acts was to acknowledge Soviet military superiority. No previous American high official had dared to call that bitter fact to public attention. Two years later, the President addressed the greatest deficiency in American military preparedness. He proposed to launch a comprehensive research effort with the aim of rendering weapons of mass destruction impotent and obsolete.

The U.S. aim has been to prevent war. The means prescribed by the policy of Mutual Assured Destruction was retaliation. The aim is still to prevent war. The new means is to establish defense and make the outcome of an attack uncertain. There is no miracle cure that will end the threat of war. The hope is to find the means of defense that in the long run will prove more effective than any further deployment of offensive weapons.

A standard argument against defense is that it is expensive. The fact is that the budget for SDI in 1985 amounted to about 1 percent of the total U.S. military appropriations, which in turn is less than 28 percent of federal spending. In the end, the balance between aggression and defense will depend on which is more effective, in other words, which requires less money. To deploy defense must cost less than to destroy it. Defense must also cost less than deployment of more weapons of attack to compensate for the missiles shot down by the defense.

If we find real cost-effective ways for defense, then each side will give most of its attention to defense. That is the aim of our research. Today we only can guess at the economic feasibility.

The President's initiative on defense so far has had two important consequences. One is that we are taking a much more diversified view of defense than ever before. We are considering the old terminal defense. We also are investigating what can be done during the midcourse, while the rockets are traveling outside the atmosphere. Can a rocket be distinguished from a decoy? Can the rockets be destroyed? We are beginning to put more emphasis on the first couple of minutes after a rocket is fired. That is the vulnerable stage. Perhaps defense against short-range rockets will become feasible in the near future. A short-range rocket will rise over the horizon while it is still accelerating and is easy to destroy or deflect.

The second result is that we have found that defense is less difficult than was generally expected. We also understand past Soviet efforts more clearly. Wherever we find "new" possibilities, we have indications that, for the Soviets, they are quite old.

That the shield is better than the sword is consistent both with what the Soviets used to say and with Russian behavior through the centuries. Only after President Reagan agreed with them did they start denying it. The intensity of their public denial is impressive.

In 1945 the United States won a war and was faced with a new peril of enormous magnitude. We lost the ocean barriers that made us secure. That security was destroyed by the genie of technology, and nothing can put the genie back into the bottle of ignorance. Many Americans do not want to settle for less than the kind of complete security we once enjoyed. That security is gone forever.

The evidence provided by our research and by observations of developments in the Soviet Union lead to one conclusion: The genie that produced the sword of modern times can also produce the shield. If we can make the shield more effective than the sword, we shall have made the decisive step that eventually will render the sword obsolete.

3

Who Is Ahead?

Will Soviet defenses make American retaliation impotent and obsolete? Are the arguments given in the preceding essay valid? Do the Soviets have an advantage?

Everyone who has visited in the Soviet Union is aware of the inadequacies of the country. The housing, even in new hotels, is substandard; most clothes and shoes are drab and ill-fitting; automobile traffic is a small fraction of that in other developed nations. Not only are the checkout counters of the stores in Moscow not computerized (as they are, for instance, in Buenos Aires), but customers must carry their handwritten chit to three different people before they can claim their purchase. Sporadic shortages make buying something as ordinary as aspirin a whole day's challenge. Meat is chronically in short supply. Shoddy workmanship, inadequate supply, and inefficiency characterize the Soviet economy. Clearly, such a country could not be a military threat.

U.S. officials and scientists who have worked with the Soviets on joint space efforts during the early 1970s can add to the dismal picture. They report that the top echelon of space program personnel is excellent, indeed outstanding, in courage, competency, and dedication. However, they add, the second rank of workers is so poor as to give any reasonable person doubts about the entire program. What the Soviets have accomplished in their space program is miraculous, considering the average quality of their workers, and has undoubtedly been achieved at the expense of failures. If a sufficient number of well-trained, knowledgeable people cannot be found for a program as important as space exploration, how can anyone believe that the technical excellence required for modern defense can be acquired?

Every fall, the Soviets stand in line to buy American wheat. With one-sixth of the world's land mass, they cannot grow enough grain to feed the 280 million people who live within their borders. Surely a country that cannot manage to fulfill the simplest requirement of well-being—enough food—cannot invent and develop a sophisticated military system.

Conversely, the Soviet scientist or performer who visits the United States is amazed at the number and size of skyscrapers, bridges, and superhighways. With so much wealth spent on nonessential things, what must the strength and depth and variety of the military defenses be like? The Americans are greatly to be feared.

We know about Soviet civilian accomplishments. They know about ours. What happens in the military field is less clear. In the absence of information one tends to extrapolate in a straightforward manner on the basis of one's own behavior. That is an extremely unreliable basis for decisions. A lot of information about the Soviets is available to our intelligence community, but it is kept secret. The facts are deeply buried. Assumptions, repeated so often that they become conclusions, take their place.

The manner in which we became aware of offensive rockets in the Soviet Union is an illuminating example. The launching of Sputnik came as a surprise. The Soviet Union was the first in space, and Khrushchev was not reticent about the accomplishment. American reaction was vigorous. The positive result was Kennedy's program to land on the moon. The negative result was the imagined missile gap, which became real by 1980.

In the late 1960s I became aware of the fact the Soviets were engaged in massive deployment of nuclear-tipped intercontinental rockets. The information was highly classified; the public remained unaware of the danger.

The secret was released in 1972. In negotiating the first SALT (Strategic Arms Limitations Treaty) agreement, the American team told the Soviets what we had found out about their missile deployment. The Soviet representatives did not deny or confirm those statements. SALT I was framed on the basis of our information. There was one clearly positive result: Congress and the American people were informed.

The one-sided flow of information is continuing. The obvious result is a lack of a clear picture about Soviet military power. Have the SALT agreements limited Soviet deployment? The answer depends more on prejudices than on facts. When Khrushchev lost power, the Soviets ceased being boastful. Today, the public knows less than ever about Soviet military strength.

A more recent example concerns the production of titanium metal in the Soviet Union. Titanium is better than steel, better than aluminum. It is abundant in nature but difficult to fabricate. If it is not melted and cast in a vacuum, it will contain bubbles and become brittle. If properly handled, it is the toughest of all available metals by weight. The Soviets use titanium for the hulls of submarines.

But one time in the post-Khrushchev age, the Soviets did boast. They erected a magnificent monument in honor of their astronauts. The monument is shown to foreign tourists. The fact that the remarkable representation of the sweeping exhaust of the rocket (which on the monument supports the rocket) is made of titanium is not usually mentioned. Nor is it said that such a monument would have required considerable effort for the United States to build. Titanium is an example of Soviet technical excellence. Other examples are kept secret.

In the 1960s the public was uninformed about Soviet offensive power. Today the public is uninformed about Soviet defensive power. There is a close connection between knowing and wanting to know. The relation is almost as close as between believing and wanting to believe. During the 1930s, the democracies of the world did not want to know about Hitler's rapidly developing military machine. Today they do not want to know about steadily growing Soviet military power. History does not repeat itself. Its parallels are incomplete and not to be trusted. But neither should they be disregarded.

4

The Antiweapons

In my high school Latin class I learned the sentence *Si vis pacem, para bellum*—If you want peace, prepare for war. Today, conventional wisdom asserts the contrary view. The assumption that underlies our long-lasting political effort toward arms limitation and disarmament is that the fewer the weapons, the greater the safety. General statements of either kind are dangerous. A brief review of history suggests the origin of the second opinion.

The wars of the twentieth century differ sharply from previous wars. After technology was applied in the military field, innovation in weapons became rapid and decisive; the effect of those innovations became unpredictable. During World War I, defense predominated; machine guns, high explosives, and trench warfare wrote their terrible pages in history. Defense at that time took only one-fourth as much effort as attack. In World War II, attack prevailed, together with weapons of mass destruction. Whether attack and mass destruction are necessarily connected is debatable. That they occurred together is a fact. World War I killed 10 million people; World War II, 50 million. The atomic bomb ended World War II. It also established the idea of mass killing as the image of any future war.

Military planning during the past forty years has been deeply influenced by the availability of nuclear explosives that could do tremendous damage at a limited cost. Together with delivery systems— either airplanes or rockets—these weapons became the measure of military power in an imagined cataclysmic war. But at the same time, in actual conflicts, smart weapons have been developed, weapons that can seek out individual targets. They play a very different role. Yet the possibility of indiscriminate annihilation remains important as an instrument of blackmail.

President Reagan named the current effort in weapons development the Strategic *Defense* Initiative. The Soviets claim it is meant to provide "an attack from the skies." What is the truth? The difference between attack and defense is enormous. But examining deployed weapons, or their software, or even the research that is going to lead to weapons, shows that sharp and verifiable distinctions between weapons of attack and defensive weapons are difficult to establish. The difference often depends on intentions at the moment, and those cannot be foreseen.

However, the purposes for which a weapon can be used are to some extent limited. It is possible to determine whether a weapon is constructed to produce mass destruction or whether it is a weapon of individual scope. Some weapons of sharply localized effect are designed for use against weapons of mass destruction. The expression *antiweapon* is an appropriate term for such weapons.

The difference between weapons and antiweapons can be argued in a clear-cut fashion. The distinction will not apply to all weapons, but extreme cases are unmistakable. A 10-megaton explosive, not very accurately aimed, is a weapon of mass destruction. A precisely directed laser beam of high intensity is an antiweapon.

Is Defense Possible?

Missiles move a few miles per second, are not easy to detect or track, and are difficult to distinguish from cheap, numerous decoys. In the 1950s those facts led the United States to the conclusion that practical defense is impossible. Today there is increasing evidence that modern technology can produce successful defense measures. Even without great innovations, effective tools are available. However, no single method will solve the problem of defense. Therefore, research must remain the highest priority.

At the same time some defensive measures are sufficiently realistic to justify deployment. Early deployment of defense has several advantages. It would provide protection against single missiles from any source. More importantly, such deployment would identify any defects in engineering. A reasonable assessment of the costs of defense cannot be provided without deployment, and, in order to assess potentials accurately, as many promising defensive weapons as possible should be deployed.

The first requirement of defense is to identify and track the approaching missiles. Radar, the usual means for that, has been greatly improved. However conventional radar installations are big, expensive, and vulnerable. One alternative is to use small, mobile equipment. A further improvement involves using laser beams to play a role similar to that of the radio beams of radar. Used in addition to radar, lasers may be used not only to detect the missiles but also to destroy them.

The most difficult task connected with defense is to distinguish a missile from a decoy. The first job is not to stop an incoming missile but to determine whether it is heavy and therefore a dangerous weapon, or lightweight and therefore a decoy. Decoys of necessity are lightweight; sending a large number of heavy objects across thousands of miles is prohibitively expensive. It is not easy to make a cheap decoy look like a heavy, substantial missile, because decoys, unlike the heavier weapons, are easily deflected from their courses. One old, practical means of discriminating missiles from decoys is to wait until the missiles reenter the atmosphere. Simple decoys will be slowed by the air 50 miles above the earth's surface. The missiles that continue a steady approach are the ones that must be destroyed.

A short-range missile never really gets out of the atmosphere, which means that decoys will not work. Therefore, defense against missiles fired from 100 to 500 miles away is likely to be the first available. The methods used against short-range projectiles can be extended gradually to missiles fired over greater distances.

One form of defense that is currently feasible involves the antimissile missile. If radar and lasers can find and track the incoming missile, then a defensive missile can be guided to shoot it down. The defensive missile must be fast and should have a range that includes the highest parts of the atmosphere. After missiles have reentered the atmosphere, they may extend small, stubby wings and fly a practically unpredictable path to a predetermined target. Such maneuvers would defeat a defensive rocket that had to make actual contact with the attacking weapon.

Defense is easier and more reliable if the antimissile missile is armed with a small nuclear explosive. A well-designed explosive, equivalent to no more than 10 tons of TNT, could destroy the incoming missile from a distance of a few hundred feet, particularly if the explosive is designed to aim its energy in the proper direction. If used within the atmosphere, such explosives must be limited in power, not only to avoid damage on the ground but also so as to cause minimal confusion in the defensive system. If the offensive missile is stopped at least a mile above the earth's surface, the defended population could probably sleep through an effective defense, even if small nuclear explosives are used.

Terminal defense, destroying the incoming missiles when they are close to their targets, is feasible with or without small nuclear explosives. The antimissile missile is only one way to destroy the attacking weapons. Once the course of the incoming missile is known, small rocks (or fragments of any material) can be placed in its path. The fast moving attack rocket will be destroyed by its collision with appropriate fragments. Such obstacles, a sort of shield, do not need to move rapidly.

At the other extreme, defensive fragments may move with velocity greater than that of the incoming missile. Conventional bullets are driven by the high-temperature gases produced by burning gunpowder. As a consequence, the velocity of bullets is limited to the velocity of the gas driving it. Transferring the energy to hydrogen, a gas that can move particularly fast, makes it possible to increase the velocity of the bullet. A more radical solution is to drive the bullet with electromagnetic forces. That is the idea behind the rail gun, which can propel objects with speeds in excess of 10 miles per second. Rail gun bullets have a finite but substantial range within the atmosphere.

After the missile has been hit, a question remains: Has it become truly ineffective? Nuclear weapons can be constructed to detonate with full force when touched by any object. A megaton explosion at an altitude of 2 miles would be disastrous. A 10-ton defensive nuclear explosion would not be noticed. It could rapidly destroy and prevent detonation of the attack weapon. A fragment of hypervelocity, traveling at 30 miles per second, could also prevent detonation. However, if the missile were carrying biological agents, the fragments would continue to be dangerous. In that case, a small nuclear explosion would be needed to destroy the bacteria or viruses and prevent widespread casualties.

Developing antiweapons to detect, discriminate, track, and destroy or neutralize offensive missiles is difficult. New technologies have made the job easier. In the end, technology will probably accomplish more than the most optimistic forecasts. That is true not only in defense but in communications, flying, medicine, and perhaps even in the most difficult of all questions, feeding the world's population.

FAST ANTIWEAPONS

The defensive weapons described so far are effective in the last stage of attack. Antiweapons that can attack the incoming missile earlier in its course may have to be deployed in orbit during peace-time. It is difficult to know how long such predeployed weapons will survive and whether they will remain effective. If a missile is to be attacked before it reenters the atmosphere or even while it is being accelerated, fast-moving antiweapons are most effective.

One of the great modern scientific discoveries is Einstein's state-ment that nothing can move faster than light—about 186,000 miles per second. Some antiweapons, the particle beam weapons, ap-proach that speed. Research on charged particles—electrons and atomic nuclei—has been conducted for many decades. When a par-ticle is accelerated, it approaches the speed of light very gradually. But while the velocity of the particle increases slowly, the energy of the particle increases appreciably. The slightly faster particle has lots more energy and will do far more damage on impact.

Antiweapons operating at velocities close to that of light have the advantage that they can be aimed at a missile more than a thou-sand miles away and hit it before the missile can change speed or direction. Missiles move with not much more than one hundred-thousandth (.00001) of the velocity of light. A fast antiweapon would need to be aimed only about a hundred feet ahead of the target, even if the target is thousands of miles away.

However, electrons, lightweight negatively charged particles, are deflected by the magnetic field of the earth. Taking that deflection into account is practically impossible. The earth's magnetic field is complex and changes with time. At no moment can the state of the magnetic field be determined throughout the path of a particle aimed at a distant point.

In the case of electrons, the problem can be overcome by pre-ceding the electron beam with a moderately intense laser beam. In the high, very dilute atmosphere, such a laser beam will ionize the few atoms and molecules it encounters. The lightweight electrons dif-fuse away, while the heavier positively charged ions remain in place. If the negatively charged beam is sent after the laser beam, the elec-tron beam will stay within the positively charged straight tunnel left by the laser beam. The earth's magnetic field is too weak to disturb such an arrangement. That method works in the upper atmosphere, where there is an appropriate density of molecules.

Instead of electrons, hydrogen atoms may be the particles used. Atoms have a neutral charge and so are unaffected by the earth's magnetic field. But particles have to carry a charge in order to be accelerated by an electric field. A hydrogen atom can weakly hold a second electron, in which case it has a negative charge and can be accelerated. When the negatively charged hydrogen ion is moving fast enough, the extra electron can be stripped away. Thus, a fast neutral atomic beam, moving in a straight line, unaffected by the magnetic field, can be produced.

The technique for straightening the path of an electron beam and the method for accelerating hydrogen atoms both come from the open Soviet scientific literature. After the initial publication, neither idea received further comment or discussion in openly circulated Soviet journals.

Electron beams and neutral particle beams could make remarkable antiweapons. Neutral particle beams, together with appropriate detection equipment, could identify the missiles among many thousands of approaching objects. The high-energy atomic beams would penetrate the rocket and emit telltale radiations. Decoys are too light to produce enough radiation to be detected. The best use of the electron beam may be to damage incoming rockets in the very high atmosphere.

Neither electron beams nor neutral particle beams will penetrate the atmosphere. Neither can be fired from the ground or used to attack objects on the earth's surface. The apparatus producing them must be predeployed or popped up on warning in a very short time.

There is no such thing as an ultimate weapon, but a laser represents the ultimate in speed. Lasers are the best antiweapons, and their development has produced the unexpected result of making defense feasible.

The process of forming a laser beam can be thought of as the opposite of casting a shadow. A beam passing through a gas that absorbs half the light in an inch of travel will decrease to one four-thousandth of its original intensity in a foot of travel and to one sixteen-millionth in 2 feet. Strange states of matter can be produced in which the opposite process occurs; energy deposited in matter can be released under appropriate conditions. In this case, a light beam may double in intensity in 1 inch, grow four-thousandfold in a foot of travel, and in 2 feet may increase 16 million times. The laser light so produced propagates strictly in the direction of the original light and tends to perpetuate all the properties of that light. It can be made extremely intense and sharply defined. No magic is involved. Energy was deposited; the laser only reemits that energy in a sharply defined direction at the proper time.

The theoretical foundation for lasers was suggested by Einstein in 1916. More than three decades later, Einstein's idea was put into practice. Many more years passed before lasers of sufficient intensity to become practical weapons were developed. (By contrast, the development of the atomic bomb took only six years.) The first actual implementation of Einstein's idea was realized in 1959. It was not applied to light but to microwaves, which have wavelengths more than a thousand times longer than light.

At appropriate intensities, microwaves may be used as antiweapons. Microwaves have an advantage over light in that they penetrate clouds with relative ease (because water droplets are small compared to their wavelength). However, microwaves are difficult to emit in a sharp direction, and high-intensity microwaves cause a discharge in the air, which dissipates their energy. Above the atmosphere, microwave lasers could be effective antiweapons over a short range.

High-intensity infrared light (with a wavelength longer than visible light) can be emitted with relative ease. Such lasers can also be quite sharply directed, provided huge mirrors are used to focus them. They have been proposed as predeployed antiweapons in space, and in appropriate weather (they penetrate clouds with difficulty) they could shoot back down to earth at limited targets. However, predeploying lasers in space is expensive, and space defenses would be vulnerable to attack by ground-based lasers, such as those already deployed by the Soviets.

It is not necessary to predeploy the lasers in space. At extremely high intensity, ground-based lasers may be used against approaching missiles. It is also possible to pop up mirrors to redirect the beam of a ground-based laser. However, for infrared lasers, such mirrors have to be quite large. Lasers producing shorter-wavelength beams (of visible light or ultraviolet light) are even more effective in damaging the target. In addition, smaller mirrors are sufficient to produce and redirect such beams.

Lasers will not penetrate clouds, because the beam is scattered by water droplets. With appropriate lasers, it is possible to destroy the droplets in a thin cloud, thereby producing a tunnel through the cloud. To have defense in all weather conditions, multiple laser locations will be needed.

The passage of light through the atmosphere is not smooth, even without clouds. Stars appear to twinkle because of the turbulence in the atmosphere. The same turbulence may destroy the sharp directionality of laser beams. The penetration of intense laser beams through the atmosphere has further limitations, partly due to heating (and subsequent distortion of the atmosphere) and partly due to peculiar types of scattering that occur with high-intensity laser beams.

The principle of very long-wavelength lasers was discovered in the United States by Charles Townes. The Nobel Prize for the discovery was shared by Townes and two Soviet scientists, Nikolai Basov and Aleksandr Prokhorov. Today, Townes is working on astronomy, including the interesting question of whether there is a black hole in the center of our galaxy. Basov and Prokhorov have continued along more practical paths, including military applications.

In 1964 Townes, together with Raymond Chiao and the Soviet physicist Boris Stoichev, discovered a phenomenon known as stimulated Brillouin scattering. It occurs when the beam of a high-energy laser passes through a transparent medium such as air or water. In 1972, the Soviet scientist Boris Zeldovich discovered that the phenomenon has practical applications to the manipulation of laser beams at high intensity. In December 1985 *Scientific American* published a beautiful article by Zeldovich explaining the principles of his discovery.[1]

This history of laser applications now covers nearly thirty years. The means of storing energy in an appropriate form so that it can cast a negative shadow requires special tricks in every case. Essentially the same effect can be produced by free electrons, atoms, molecules, crystals, or semiconductors. The art is as variable and tricky as chemistry itself.

Recently the art has been extended to x-ray lasers, which can work at wavelengths a thousand times shorter than visible light. The x-rays are most effective in damaging a target. One way to produce an x-ray laser is to use the energy from a nuclear explosion, which can be stored for an exceedingly short time in highly ionized atoms. Using an appropriate nuclear explosive to produce sharply directed x-ray laser light is truly a novel art. The suggestion again came from Soviet scientific publications.

The x-ray laser may have tremendous consequences in defense, or it may have practically none. The great uncertainty is inherent in the novelty of the subject. The x-rays will not penetrate the atmosphere, so an x-ray laser can be effective only in space. That means the laser, together with its energy source, would have to be popped up on warning of attack or else may be predeployed.

Defense against missiles need not be deployed in space, but space deployment is one way to make it possible to attack a missile just after it is fired, when it is accelerating and most vulnerable. The antiweapon may well be a small, carefully guided rocket. However, predeployed defenses, which stay in orbit for years, are expensive to launch and may be relatively easy to destroy. Predeployment in space might become attractive if and when the cost of launching satellites is greatly reduced. That would, of course, have important consequences for space exploration as well as for military applications of every kind.

OBSERVATION OF THE LAUNCH

Rocket launchings are accompanied by the sudden appearance of powerful infrared radiation, which can be easily detected from space observation satellites, for instance, from a synchronous-orbit satellite that seems to remain suspended over a particular spot on earth. Space satellites, however, are vulnerable to attack. Protecting them is possible but difficult. Satellites can be hardened or made replaceable. Inconspicuous "sleeping" detectors (which would unfold and become active only if the functioning detector failed) can be deployed. The sleepers in turn can be protected by decoys.

Observation satellites can be placed in elliptical orbits, which allow them to spend most of their time in a position from which they can survey most of the Northern Hemisphere. In an orbit some thousands of miles above the earth, they would observe less detail but would still be able to serve the limited purpose of identifying a rocket launch. In such a position, they would also be better protected from attack by ground-based rockets.

Unfortunately, in most cases a satellite will not continue in the same elliptical orbit. The most distant point (the apogee) moves from north to south and back again. That can be avoided by choosing an appropriate orbital inclination (about 63 degrees to the Equator). In this *Molniya* orbit, the apogee stays put. The Soviets launched such a satellite about twenty years ago. All of their subsequent communications and their observation satellites have been placed in *Molniya* orbits.

Observation satellites should repeat their elliptic paths every twenty-four hours. Their course need not carry them over the territory to be observed, as they could see missiles emerging from the atmosphere. In such circumstances, they would be more difficult targets than satellites in synchronous orbit.

Launch detectors in space may in turn be backed up by earth-based radars. If the radars use long-wavelengths, the beam will follow the curvature of the earth over long distances. (Ionization in the high layers of the atmosphere confines the propagation of the waves.) Such a long distance detection system is called over-the-horizon radar, or OTH. It can detect the disturbances caused by a rocket launching and would provide a warning of a massive attack.

A Comprehensive Defensive System

Even assuming that a secure early warning and identification system is established and that effective antiweapons are available, the problem of defense is not yet settled. Making the decisions required to counter an attack by hundreds of missiles and hundreds of thousands of decoys within five to thirty minutes seems almost impossible. The problem of battle management appears superhuman. Neither a single person nor even a large group of highly skilled people can possibly make the fast decisions that would be necessary. Computers are required.

Deterrence by retaliation presents a far greater problem in decision-making. A mistaken decision to retaliate could trigger a horrible war and cost millions of lives. If a ballistic missile defense system were fired in error, it would hurt no one. If the missiles were not there, the only result would be wasted effort. Replacing retaliation with defense changes the problem of battle management from truly horrible to merely difficult.

Computers can make decisions with the required speed. All relevant information can be included in their programs, and their responses can be predetermined after careful consideration, including ample criticism and review. At present, no computer is fully adequate for the task of battle management. There are computers in existence, however, whose functions are comparable in complexity to those needed for defense, for example, the telephone system computer programmed with 50 million entries.

A defense-management computer would have to be more reliable than that of the telephone system. Failure in the midst of an attack has very different consequences from having a conversation interrupted. The criticism can be carried even further. Nobody is shooting at the telephone system, but the defense computers themselves will be under attack. One solution is to duplicate the computers. A fraction of the deployed computers should suffice for the management of defense.

Yet a computing system can get hopelessly confused, especially if it is partially disabled. Fortunately, it is easy to introduce self-criticism and self-surveillance into a computer. If the computer notices that it is confused, it can reduce its functions to a simpler level. The defense then would not vanish completely.

In describing the antiweapons, I have emphasized those that are novel and surprising. The planned or possible antiweapons are more numerous. However, as is true of computers, none of our instruments of defense as envisaged now are anywhere near perfection. For the time being, we must aim at establishing *some* defense, not a perfect defense.

As long as a threat of war exists, the application of technology to warfare cannot be eliminated. For the past forty years, the weapons of mass destruction have proved to be at the developing edge of science and technology. The situation today is different, and weapons that could reverse the unfortunate trend of increasing widespread destruction are now possible. It is extremely important to consider the individual antiweapons in detail before deciding whether and to what extent they should be developed and relied upon.

Defensive arms by themselves will not furnish insurance against war, nor will they terminate the arms race. But to the extent that the arms race is won by instruments of defense, the psychological conditions required for international cooperation and for developing the foundations of real peace will be approached.

5

Peace by Agreement

In early November 1985, I was among a few dozen people who participated in an unpublicized meeting with President Reagan. He repeated his proposal to deter war by defense rather than by retaliation, since retaliation would lead to the nightmare of mutual destruction. He also emphasized in an unambiguous way that when we find adequate defenses, our results should be shared with everyone, including the Soviets.

Worldwide cooperation has been suggested by the President both in public and in private. To many people, the idea appears more a puzzle than a concrete suggestion. The puzzle deserves to be addressed.

Technology has produced weapons of mass destruction. Technology may provide some safety through defense. But technology will never create an absolute weapon, either for attack or for defense. Peace will ultimately depend on agreement. Agreement between nations in turn is shaped by what weapons and safeguards technology has made available. Absolute reliance on the human element, on trust, is no more realistic than absolute reliance on the sword or on the shield. Any novel technical situation makes a new look at appropriate international agreements necessary.

Defense is now a reality in the Soviet Union; in the near future it may also be a reality in the free world. A changed approach to international negotiations has become unavoidable. But what does President Reagan mean when he says we will share protective defense with the Soviets? Obviously, democratic nations should work together toward mutual defense. That realistic approach has already been implemented with Great Britain, Germany, Israel, Italy and Japan. More nations may join the project. They have been invited to participate on practically any terms they choose.

The Soviets, however, are not interested in cooperating in de-

37

fense, even to the extent of sharing defense with the Warsaw bloc. Why should they be? Their defense program is a quarter of a century old. As the leaders in defensive technology, they have nothing to learn from others. They are probably concerned that they may lose what is now a Soviet monopoly.

In early 1967, the newly elected Governor of California, Ronald Reagan, visited the weapons laboratory in Livermore and heard about antiballistic missiles. An antiballistic missile is designed to destroy incoming rockets as they reenter the atmosphere and are approaching their targets. It seems unreasonable to wait with defensive measures until the last moment. But it is difficult to distinguish weapons from cheap and plentiful decoys in space. Such discrimination occurs naturally when the heavy weapons and lightweight decoys reenter the atmosphere. Only heavy objects continue to travel at an unchanged speed. The expensive part of an attack is the transporting of heavy objects from one side of the globe to the other. Decoys work well in space, but they fail in the atmosphere. That is why early ideas on defense concentrated on the last moments of the rocket attack. Such was the type of defense that Governor Reagan heard described, that the United States abandoned five years later, and that the Soviets have perfected and established around Moscow.

Present research on defense is more comprehensive. The Strategic Defense Initiative begun in 1983 emphasizes defense against rockets in all phases, but most particularly defense against rockets in the boost stage, the few tens of seconds after firing when the rockets must be accelerated. During that period the rockets, still carrying what will later be independent weapons and decoys, are quite vulnerable. A small, partial failure can easily lead to the destruction of the entire rocket, as was shown by the tragic Challenger accident in January 1986.

Defense against rockets in the boost phase has obvious advantages. It also has obvious difficulties. The curvature of the earth itself interferes with seeing, much less hitting, a rocket during the boost period. To stop an attack in the boost phase, one might use defensive weapons in space. That is how President Reagan's proposal was generally understood. The possibility exists and must be carefully studied, but it is not the only solution or even the best. An alternative would be to launch an instrument of defense a few seconds after the launching of a rocket. After both objects have left the dense portion of the atmosphere, the defensive weapon can be aimed at the offensive rocket from a distance of 1,000 miles. At that altitude and distance, the curvature of the earth does not present a serious obstacle.

Such a pop-up defense would be effective against medium-range missiles, those fired 1,000 miles or less from the defensive installation. It would not work if the defenses were located in the Western hemisphere and the rockets were fired from the middle of Asia. It might work from a ship or submarine located off the Atlantic or Pacific coast against rockets launched from any point in North America. Appropriate research is needed, if for no other reason, simply to determine whether American retaliation can be effective in such circumstances. Unfortunately, the same countermeasures against Soviet rockets are more difficult because of the size of the land mass of the Soviet Union.

President Reagan has emphasized the international aspect of his proposal from the beginning of his advocacy of defense. He is acutely aware of the historical importance of his office, as were Lincoln and Franklin Delano Roosevelt. An obvious and valid part of Reagan's motivation is the simple fact that it is in everyone's interest to avoid a nuclear war.

Almost half a century ago, Leo Szilard made a remark that I have not forgotten. As in many other cases, his statement was far from obvious; it was also provocative. He said that no important decision is ever taken unless there are at least two independent reasons for the action. If Szilard was right, it would seem reasonable that President Reagan may have had two very different reasons to insist on international cooperation.

One motive is the good of mankind. The other is that only by international cooperation can defensive measures succeed in preventing a massive rocket attack. President Reagan called for technology to serve the purposes of defense, and he proposed international cooperation. The latter may be more than just desirable; it may be the only way to secure the safety of everyone.

It is not and cannot be my purpose to prove that this is what the President had in mind. Even less can I show that such a proposal will succeed. But if the very promising technology of defense against missiles in the boost phase becomes a reality, the nature of negotiations would be directly affected. The present situation then might be converted to one in which stability has been achieved by mutual consent, even without a negotiated agreement on difficult questions like capitalism versus communism or disarmament and verification.

If differences and tensions persist but a consensus develops on the need for defense, will it not then become obvious that such consensus was the only solution to the problem of a threatened mutual suicide? It is a common feature of all real and essential change that looked at in prospect, it appears impossible, but looked at in retrospect, it seems unavoidable.

How do we get from here to there, assuming that the new possibilities may someday exist? It is not enough to have a goal. One also must have a road. Everyone agrees that inventions alone will not suffice. A technical fix does not work unless it is accompanied by real human will. For that reason, continued discussions with the Soviets appear a necessity. The aim of discussion is peace. Discussion might include the specific topic of preventing a rocket attack.

Up to now, negotiations have emphasized disarmament. Peace treaties and disarmament have a long history. The history of the failure of those attempts is equally long. It is possible to believe that Czar Nicholas II sincerely believed in peace through treaties when he organized the First Hague Conference on arms limitation in 1899 while Russia was arming. It is easier to see the difficulties and eventual futility of the negotiations before World War I and World War II than it is to have a realistic evaluation of the negotiations between the superpowers.

There is widespread agreement that even in the absence of success, negotiations must continue. We may indeed continue to talk but change the subject. The common feature of past and present negotiations is that in all cases they have emphasized the negative aspects. Prohibition of arms requires that mutual confidence be established between the partners to the agreement. By contrast, affirmative agreement can be reached about measures of defense. In that case, confidence is less important. The proposal is to help everyone to establish defenses. The presence of actions to provide defense would be evident. The absence of weapons, particularly in a closed society, is more difficult to prove or believe.

One possible approach might be to seek an agreement to announce all rocket launches a few hours before they occur. If the Soviets agreed to such announcement, a real beginning toward the purpose of lifting the threat of mass destruction would have been accomplished. If the Soviets say no, they will have great difficulty in maintaining that they are against the militarization of space. At the same time, we could turn to all other nations and get as wide an agreement as possible on defense against rocket attack. The system of defense against massive rocket launchings could then be deployed in as many places as possible.

In order for the system to operate in an optimal manner, the Soviet Union should join in announcing rocket launchings and eventually in preventing simultaneous launching of many rockets by anyone. The President made that point in general terms when he suggested that defensive weapons would be shared with the Soviets. Is it completely impossible that they will join in the endeavor? If China were to participate in an international agreement to prohibit massive rocket launchings, the Soviets would have a great incentive to join also. If Soviet misgivings about China are real, then the Soviets' cooperation in international defense would seem to be an obvious way to improve their own safety. With or without Soviet participation, widely deployed defensive weapons would lower the probability of a widespread nuclear missile attack.

The road to a more stable future may develop in a different manner. Some other type of agreement focusing on positive actions to diminish the possibility of a full-scale rocket attack may prove better. The suggestion of launch announcement provides only one example of an initial step toward cooperation and security.

Everyone is aware of the way in which fear of a catastrophic conflict could actually lead to a conflict. There is also a way in which fear could lead to general security. Indeed, the spectacle of one country destroying another will necessarily result in worries everywhere that the destructive power may be turned against one's own people. One may argue that bystanders in a nuclear conflict cannot possibly remain uninterested. Therefore, they should be actively involved in preventing the conflict. In an ongoing attack including long-range rockets, no one can know against whom the next salvo will be directed. All nations have an incentive to unite against any mass launching of rockets.

That part of self-interest that remains after all the details of current conditions change may properly be called morality. In that sense, it is right to talk of the interests of mankind rather than the interests of a single nation. One of the consequences of a rapidly changing technology is that it forces us to recognize that we cannot reach safety except through cooperation.

I have suggested that President Reagan had two independent reasons to propose international cooperation. One was the good of mankind; the other was the circumstance that only through cooperation can we accomplish our purpose of defense. It may be that those two purposes are not as different as they appear to be.

II

ATOMS
AND
INDIVIDUALS

6

The Story of the Atomic Bomb

The story starts at the beginning of the twentieth century in Canada, where Professor Ernest Rutherford, originally from New Zealand, was teaching at McGill University. He got himself into trouble by stating that elements are transmuted by the process of radioactivity. A chemistry professor denounced him as a fool, ignorant of the elementary fact that elements are immutable. Rutherford did not mend his ways. He continued his pursuits in England, where he performed the basic experiments that led to the understanding of the structure of the atom. He became the head of the famous institution that was the cradle of atomic energy: the Cavendish Laboratory.

I never met Lord Rutherford, but in 1935, while I was at City College in London, I heard him speak. He sounded incensed and indignant as he spoke of the suggestion that the great amounts of energy stored in the atomic nucleus might be put to practical use. He denounced as complete fools those who would believe that, because the energy contained in the innermost part of the atom is protected by the strong electric repulsion between positively charged nuclei. I did not know the reason for his indignation then, but I happened to find out a few weeks later.

Shortly before that speech, Leo Szilard had gone to Rutherford and tried to explain that the energy of a nucleus might be released. He did not get past a few sentences. Rutherford threw him out of his office. Thus the generations change. The young genius from New Zealand turned into the omniscient Lord of the hallowed Cavendish Laboratory. I in turn am probably full of old and outworn ideas. Unfortunately, I do not know which ones they are.

What Szilard tried unsuccessfully to tell Lord Rutherford was that one building block of the nuclei, the neutron, can approach any nucleus, because it has no charge. Unfortunately, neutrons do not occur in nature in the free state, and it is extremely difficult to make

45

a neutron. However, if a neutron hits a nucleus, more energy is released than a million chemical reactions could produce. Szilard speculated in a fairly detailed way about how neutrons could not only be produced but also could reproduce. He arrived at the idea that a sizable amount of beryllium, thorium, or uranium might allow a chain reaction release of neutrons.[1] Because Szilard was piqued by Lord Rutherford's rejection, he filed a patent on his idea in 1935.

In December 1938 uranium fission was discovered in Berlin. It was caused by neutrons. Would it produce more neutrons? Szilard was one of the few who promptly tried to find out. Within a few weeks, he produced a very elegant answer. Neutrons when first emitted in the fission process move at high velocity. They can be slowed down by repeated collisions with the nuclei of hydrogen atoms. Szilard bombarded uranium with slow neutrons and discovered that, along with fission, fast neutrons were produced. They must have originated in the fission process.

In March 1939, a friend and I were playing a Mozart violin sonata in my living room when we were interrupted by a long distance phone call from Leo Szilard. He said only: ''I have found the neutrons.'' To me, that meant the last doubts about the feasibility of harnessing nuclear energy had vanished.

Until then, my work and my life had been simple and straightforward. I worked in science because doing so was the greatest reward I could imagine. All other meaningful activities seemed to require accepting contradictions and compromises. In science alone can one hope to find answers that are both clear-cut and reliable.

When I learned that atomic explosives were possible, I could see that, in the event of war, no part of the world would remain safe. Half a year later, World War II began. For me, it became impossible to ignore the impact of scientific discovery on everyone everywhere. From that time onward, I was no longer a carefree scientist considering the puzzles of the universe; to an increasing extent, I worked on the applications of science. The change was not sudden and was certainly not the result of a simple, conscious choice. It was a slow-moving involvement that, in retrospect, seems to have been unavoidable.

For me, the man of the year 1939 was Leo Szilard. He drew me out of the world of science by introducing me to a world of action. However, he was not the only scientist considering the problems of nuclear energy. Niels Bohr and Johnny Wheeler had been at work. They noticed that a rare isotope of uranium, U-235, was responsible for most fissions. To separate U-235 from the common isotope, U-238, would require a massive effort.

Too massive, Bohr declared. "You cannot make an atomic bomb unless you turn the United States into a huge factory," he told us. The most imaginative man we knew would no longer imagine. He did not want to. He was happy to find a reason why such a weapon could not be produced. Szilard and Eugene Wigner, another Hungarian physicist, were no less appalled by the possibility, but they disagreed with Bohr. The possibilities of the universe do not necessarily conform to human desires. Hitler as the sole possessor of such a weapon was even worse to imagine. Those two were to a great extent instrumental in getting the U.S. atomic project started at an early time.

Eugene Wigner is a very thorough man of great imagination. He is also patient and gentle, as an incident predating the terrible period of the Nazi regime illustrates. Wigner and an astronomer friend, Otto Heckmann, were sitting on the grass at the Göttingen municipal swimming pool. Heckmann noticed a row of ants crawling across Wigner's leg and asked, "Aren't they biting?" Wigner responded, "Yes." "Then why," Heckman asked, "don't you kill them?" Wigner replied with an apologetic smile, "I don't know which one did it."

Szilard's disposition was almost the opposite: decisive, outspoken, and impatient. When he arrived at a conclusion about a proper course of action, he allowed nothing to stop him. (I should also note that he was very often correct.) Szilard and Enrico Fermi were working together at Columbia. A few months after fission was discovered, Fermi (a resident in the States for only a few months, a quiet, reticent man) had approached the Naval Research Laboratory with a request for funds for uranium research. He was turned down.

Szilard and Fermi had begun joint work on constructing the equipment needed to demonstrate a controlled chain reaction, a first reasonable exploratory step. Unfortunately, Szilard's temperament and Fermi's disposition were quite incompatible. While I was fond of both of them, they found little likable in each other besides expertise. I spent the summer of 1939 at Columbia, doing a little lecturing to graduate students but primarily as the peacemaker on the Fermi–Szilard project. They did agree, however, that graphite would be the best substance to slow down the neutrons and prepare them for fission in U-235. However, the expenses of their planned experiment were beyond their university resources.

In keeping with his temperament, Szilard was not discouraged by repeated bureaucratic rejection. He only became more convinced of the need to get governmental support of atomic research. Since direct action was blocked, he would try indirection. He conceived a complicated, effective, and completely logical idea to start the uranium project. Szilard had a banker friend, Alexander Sachs, who knew Franklin D. Roosevelt and therefore could be the intermediary in getting a letter directly to the President. Szilard decided to go to the most famous of the physicists, Albert Einstein, and persuade him to write to the President.

There was one thing Szilard could not do: He never learned to drive a car. He was perhaps the only person who has ever considered me a good driver, and he asked me to drive him to Einstein's summer home on Long Island with a typed letter, which I suspect was drafted by Einstein, Szilard, Wigner, and possibly Sachs. From its opening curt salutation to its concluding instructions on exactly what should be done, Einstein's letter sounds a little like Szilard.

> Albert Einstein
> Old Grove Road
> Nassau Point
> Peconic, Long Island
> August 2, 1939

F. D. Roosevelt
President of the United States
White House
Washington, D.C.

Sir:

Some recent work by E. Fermi and L. Szilard, which has been communicated to me in manuscript, leads me to expect that the element uranium may be turned into a new and important source of energy in the immediate future. Certain aspects of the situation which has arisen seem to call for watchfulness and, if necessary, quick action on the part of the Administration. I believe, therefore that it is my duty to bring to your attention the following facts and recommendations.

In the course of the last four months it has been made probable—through the work of Joliot in France as well as Fermi and Szilard in America—that it may become possible to set up a nuclear chain reaction in a large mass of uranium, by which vast amounts of power and large quantities of new radium-like elements would be generated. Now it appears almost certain that this could be achieved in the immediate future.

This new phenomenon would also lead to the construction of bombs, and it is conceivable—though much less certain—that extremely powerful bombs of a new type may thus be constructed. A single bomb of this type, carried by boat and exploded in a port, might very well destroy the whole port together with some of the surrounding territory. However, such bombs might very well prove to be too heavy for transportation by air.

The United States has only very poor ores of uranium in moderate quantities. There is some good ore in Canada and the former Czechoslovakia, while the most important source of uranium is the Belgian Congo.

In view of this situation you may think it desirable to have some permanent contact maintained between the Administration and the group of physicists working on chain reaction in America. One possible way of achieving this might be for you to entrust with this task a person who has your confidence and who could perhaps serve in an unofficial capacity. His task might comprise the following:

a) to approach Government departments, keep them informed of the further development, and put forward recommendations for Government action, giving particular attention to the problem of securing a supply of uranium ore for the United States.

b) to speed up the experimental work which is at present being carried on within the limits of the budgets of university laboratories, by providing funds if such be required, through his contacts with private persons who are willing to make contributions for this cause, and perhaps also by obtaining the cooperation of industrial laboratories which have the necessary equipment.

I understand that Germany has actually stopped the sale of uranium from the Czechoslovakian mines which she has taken over. That she should have taken such early action might perhaps be understood on the ground that the son of the German Under-Secretary of State, von Weizsaecker, is attached to the Kaiser-Wilhelm-Institute in Berlin where some of the American work on uranium is now being repeated.

Yours very truly,
Albert Einstein

Finding Einstein's house was not easy. We finally asked a young girl (perhaps eight years old) where the man with long white hair lived, and she pointed out the house. Afterward, while the three of us were drinking tea together, Einstein mused: "Yes. This would be the first time on earth . . . " The possibility of employing nuclear energy did indeed seem close at hand.

Einstein signed the letter on August 2, before the invasion of Poland. Roosevelt received it after Poland was defeated. He called Lyman Briggs, the head of the Bureau of Standards. Briggs called a meeting to which several people, including Enrico Fermi and me, were invited. Fermi was not optimistic about the possibility of research funds. At Szilard's urging I saw him and attempted to persuade him to attend. Fermi, still unhappy over his prior brush with bureaucracy, refused. He asked me to represent him and present his views. From chauffeur I had been promoted to messenger boy.

The meeting was held in November, and Szilard and Wigner presented the case. A representative of the Army, a colonel whose name I have forgotten, then commented: "At the Aberdeen weapons testing grounds, we have a goat tethered to a post with a ten-foot rope. We have promised $10,000 to anyone who can kill it with a death ray [today known as a laser], but nobody has yet claimed the prize. I don't believe any of the foolishness you scientists are talking about death rays or atomic bombs. Furthermore, wars are not won by weapons. They are won by moral superiority."

My turn to speak (for Fermi) came at the end of the colonel's comments. I merely reiterated Fermi's belief that research on atomic explosives should start by producing a chain reaction. That step would require exceptionally pure graphite to slow down the neutrons. To secure the graphite needed for the first year of research would probably require $6,000. The sum was quickly granted.

Wigner was horrified by, and Szilard furious at, this grossly inadequate budget proposal. By the summer of 1945, the sum of $2 billion had been spent.

I was uncertain as to whether I wanted to be a bystander or a participant in the project. I was, as I had dreamed of being, a professor. Because of the Nazi revolution, I had moved from Göttingen to Washington, D.C., from a familiar to an unfamiliar ivory tower. But it was still an ivory tower. Physics was certainly a more interesting activity than politics. Yet, while I was happy in academia, the speedy fall of Poland and the possible new weapon made me worry even more about how political decisions might affect the future.

I continued my usual routine at George Washington University through the spring of 1940. About mid-April, I was invited to the Eighth Pan American Scientific Congress, which was to be held on May 10. President Franklin Roosevelt was to give the keynote address. I was somewhat curious about Roosevelt, but I did not intend to go. Then, on May 8, the Nazis began rolling through Holland, Belgium, and Luxembourg. I decided to use my invitation.

Roosevelt's speech that day described how long it would take for a German bomber to attack the various capitals of Europe. The President also talked about the right of small nations to exist and determine their own course. However, Roosevelt really caught my attention when, referring to the Blitzkrieg (which forty-three days later would leave the Allies without a toehold on Continental Europe), he said:

> You who are scientists may have been told that you are in part responsible for the debacle of today because of the processes of invention for the annihilation of time and space, but I assure you it is not the scientists of the world who are responsible, because the objectives which you held have looked toward closer and more peaceful relations between all nations through the spirit of cooperation and the interchange of knowledge. . . .
>
> The great achievements of science and even of art can be used in one way or another, to destroy as well as create. They are only instruments by which men try to do the things they most want to do. . . .
>
> Can we continue our peaceful construction if all other continents embrace by preference or by compulsion a wholly different principle of life? No, I think not.
>
> Surely it is time . . . to use every knowledge, every science we possess, to apply common sense and above all to act with unanimity and singleness of purpose.
>
> I am a pacifist. You, my fellow citizens of twenty-one American republics, are pacifists too.
>
> But I believe that . . . you and I, if in the long run it be necessary, will act together to protect and defend by every means at our command, our science, our culture, our American freedom and our civilization.

Seated in the crowd that day, knowing of Einstein's letter and the Briggs committee, I thought Roosevelt might be thinking about the highly probable, immensely powerful new weapon as well as about the Blitzkreig instruments. In fact, I felt almost certain that the new knowledge was uppermost in his mind.

But more important, so pertinent were his remarks to the questions I was struggling with, I had the strange impression that he perhaps was talking directly to me. He was suggesting not something that scientists might do, but something that it was our duty to do— to use our abilities to protect and defend what we valued.

Roosevelt spoke for twenty minutes. My mind was made up. I have never regretted my decision. What is frightening to me is that today, with a different, less crudely posed threat but one fully as dangerous, Reagan's similar appeal is not heeded by many scientists.

In March 1941, Mici and I fulfilled the residency requirement and, sponsored by Merv Tuve and Ferdinand Brickwedde, were sworn in as citizens. About the same time, we moved into our first home. We had hardly settled in when I was asked to come to Columbia as a visiting professor for the following year. It appeared on the face of things that I would be going into a similar teaching role, but it was clear to me that the post was offered primarily because I was needed on the atomic reactor project.

The real work for me started in 1943, when Oppenheimer established the nuclear explosives laboratory at Los Alamos. At first the work was not so different from what I had been doing and what I was trained to do. We needed new ideas, and we needed to consider them carefully and critically. One of the first important questions that arose in Los Alamos concerned predetonations. Although neutrons caused fission in a finite mass of uranium (or its derivative, plutonium) and fission produced more neutrons, the result would not necessarily be a large explosion. If neutrons escaped, the energy release would occur at a steady, nonexplosive level. If a critical amount is present, more and more energy is set free, but expansion of the material stops the reaction before much happens. That is called predetonation. A substantial, supercritical amount is needed for an effective explosion. That means nuclear materials must be brought together at high speed.

A young man at Los Alamos, Seth Neddermeyer, found the best solution: implosion. Let high explosives drive a spherical shell of uranium or plutonium together toward its center. Would it work? It had to be done with great symmetry; the answer was in doubt. Two months after the Los Alamos laboratory was started, Johnny von Neumann arrived at the lab. His first evening there, he came home with me to have dinner. My old friend from Budapest was a truly great mathematician with a practical mind. When I began telling him about Neddermeyer's implosion method, he became very interested in thinking through the problem in detail.

Solids are incompressible under ordinary pressure. But as we discussed implosion, I realized from my earlier work on the structure of matter that under imploding conditions, the uranium would be strongly compressed. The increased density of the uranium would mean that neutrons could not escape as easily. Therefore, the amount of material needed for a critical mass would be much smaller. Making the fuel for explosives was an extremely slow and painstaking process at that time. The newly recognized implosion effect made it feasible to produce a bomb in time to affect the war.

The next day Johnny and I presented the idea to Oppenheimer. He accepted it at once. I have yet to see another organization that could move as fast in the right direction as Oppenheimer's laboratory.

Then the troubles started. The general ideas were easy. The execution was difficult. Considering the limited time available, the planned calculations might have been impossible had Johnny not introduced IBM computers.[2] They were incredibly slow. Today's electronic brains move a billion times faster. My friend Hans Bethe was in charge of theory and computations. He wanted me to take hold of the execution. I refused. I knew there were others better at it. I also wanted to pursue more novel ideas, including fusion.

There is a further twist to the story. Bethe's group completed the elaborate calculations, but nobody would believe them, not even the people who had performed them. In the end, a much simpler design was adopted for the plutonium bomb used against Japan. The calculations needed for it could almost have been carried out on the back of an envelope.

The result of our wartime work at Los Alamos was tested in southern New Mexico on July 15, 1945. As the oldest member of our group (I was thirty-seven), I was invited to see the test from an observation area just 20 miles away. I wanted to view the shot directly, but we all had been instructed about the dangers of the heat and light of the explosion. Therefore, I put on gloves, covered my face with suntan lotion, wore dark glasses, and held the prescribed piece of welder's glass to my face. I stared straight at the spot as the countdown began. The numbers descended steadily from thirty down to five. I then waited through what seemed a very long period of silence. I felt sure that the test had failed. Finally, a faint point of light appeared and seemed to separate into three parts.

As soon as my brain started to register disappointment, I remembered my double eyeshields. I eased up the welder's glass, looking down at the ground. It was as if someone had opened the curtains of a completely black room to a flood of sunlight. I was impressed.

We were taken back to Los Alamos by bus. I went home. My wife did not ask me where I spent the night. She said: "I heard on the radio that an ammunition dump blew up, but no one was hurt." I did not comment. I tried to sleep. After a few hours, I went to the office. Just as I sat down, a young woman from our group, Mary Argo, rushed in crying: "Mr. Teller, Mr. Teller, have you ever seen such a thing in your life?"

Mary had been part of a group of young people that had hiked to the top of Sandia Mountain, about 70 miles from the test site, the night before in order to watch the test. The shot had been delayed, a fact they did not know. They waited an hour before giving up in disappointment. As they started down, the sky was lit up by the first atomic explosion.[3] In Los Alamos my metamorphosis was completed. In January 1939 I had been a pure theoretical physicist. Before the attack on Hiroshima I had started work in applied science. After the war, I tried to find my way back to the simpler life of a scientist and a teacher.

I never succeeded.

7

Seven Hours of Reminiscences

When Shakespeare presented the life story of Henry VI, he wrote it in three parts, and the plays in their entirety ran more than seven hours. A BBC drama compressed the life of J. Robert Oppenheimer into seven hours, a considerable accomplishment, as to my mind Henry VI was not nearly as extraordinary, ingenious, or self-contradictory a character as Oppenheimer. Most of us probably imagine ourselves and our close associates to be simpler than we are. However, the complexity of the man I knew and worked with makes the television representation seem almost one-dimensional.

The film indicates but does not show the contrast of the historical poles of Oppenheimer's life—his work at Los Alamos and his loss of security clearance. Oppenheimer's accomplishments as war-time director of the Los Alamos Laboratory are unique. Much of my life has been spent in laboratories of similar size and nature. I have known many laboratory directors intimately. For a short time, I was even a director myself. I know of no one whose work begins to compare in excellence with Oppenheimer's.

Throughout the war years, Oppie knew in detail what was going on in every part of the laboratory. He was incredibly quick and perceptive in understanding human as well as technical problems. Of the more than ten thousand people who eventually came to work at Los Alamos, Oppie knew several hundred intimately. He knew their relationships to each other. He knew what made them tick. He knew how to organize, cajole, humor, soothe feelings—how to lead powerfully without seeming to do so. He was an exemplar of dedication, a hero who never appeared to be one and never lost his humanness. Disappointing him carried with it a sense of wrongdoing. Los Alamos's success grew out of the brilliance, enthusiasm, and charisma with which Oppenheimer led it.

I became aware of a different side of Oppie in June 1945, a couple of weeks before the Alamogordo test of the first atomic bomb. I had received a letter from my good friend Leo Szilard.[1] It contained a petition from Chicago. Szilard suggested that I sign and circulate it among my colleagues at Los Alamos:

July 4, 1945

Dear

Inclosed is the text of a petition which will be submitted to the President of the United States. As you will see, this petition is based on purely moral considerations.

It may very well be that the decision of the President whether or not to use atomic bombs in the war against Japan will largely be based on considerations of expediency. On the basis of expediency, many arguments could be put forward both for and against our use of atomic bombs against Japan. Such arguments could be considered only within the framework of a thorough analysis of the situation which will face the United States after this war and it was felt that no useful purpose would be served by considering arguments of expediency in a short petition.

However small the chance might be that our petition may influence the course of events, I personally feel that it would be a matter of importance if a large number of scientists who have worked in this field went clearly and unmistakably on record as to their opposition on moral grounds to the use of these bombs in the present phase of the war.

Many of us are inclined to say that individual Germans share the guilt for the acts which Germany committed during this war because they did not raise their voices in protest against those acts. Their defense that their protests would have been of no avail hardly seems acceptable even though these Germans could not have protested without running risks to life and liberty. We are in a position to raise our voices without incurring any such risks even though we might incur the displeasure of some of those who are at present in charge of controlling the work on "atomic power."

The fact that the people of the United States are unaware of the choice which faces us increases our responsibility in this matter since those who have worked on "atomic power" represent a sample of the population and they alone are in a position to form an opinion and declare their stand.

Anyone who might wish to go on record by signing the petition ought to have an opportunity to do so and, therefore, it would be appreciated if you could give every member of your group an opportunity for signing.

Leo Szilard

The Chicago laboratory, headed by Arthur Compton, had worked on planning the production of material for the bomb. Their work had been completed some months earlier, and Szilard, James Franck (nicknamed Pa Franck), and several scientists in the project had some time to consider the political and moral issues related to the bomb, a development I knew about because I had recently been in Chicago. The petition they drew up, addressed to the President, pointed out that scientists began work on the atomic bomb because we might have been attacked by this means, but that danger had ended with Hitler's suicide. It noted that the ruthless destruction of cities would be added to further if the bomb were used, which would set a precedent and open "the door to an era on devastation of an unimaginable scale." The petition asked the President "to rule that the United States shall not, in the present phase of the war, resort to the use of the atomic bombs."

I was inclined to sign the Chicago petition, but I could not circulate it at Los Alamos without checking the matter with Oppenheimer. I arranged to talk with him at his office. The BBC film suggests that other people accompanied me, but only Oppie and I were present at this conversation. I showed him the petition.

Oppenheimer immediately offered several uncomplimentary comments about the attitudes of the involved Chicago scientists in general and of Szilard in particular. He went on to say that scientists had no right to use their prestige to try to influence political decisions. He assured me that the right decisions would be made by the leaders in Washington, who were wise people and understood the psychology of the Japanese. I have a vague recollection that he referred to George Marshall as an example of proper leadership. My predominant feeling following our conversation was relief—I did not have to take any action on a matter as difficult as deciding how the bomb should be employed.

Later I learned that shortly before that interview Oppenheimer not only had used his scientific stature to give political advice in favor of immediate bombing but also had put his point of view forward so effectively that he gained the reluctant concurrence of his colleagues. Yet he denied Szilard, a scientist of lesser influence, all justification for expressing his opinion.

In fact, four scientists were asked to serve as an advisory panel on the use of the bomb: Arthur Compton, from whose laboratory the petition originated; Ernest Lawrence from the isotope separation laboratory at Berkeley; Oppenheimer from Los Alamos; and Enrico

Fermi, whose political discretion was carried to the point of hardly ever expressing an opinion that differed from the majority.[2] Only Fermi and Oppenheimer were aware of the mechanics and expected effects of the bomb itself. Only Oppenheimer advocated immediate use of the bomb.

The flow of information between laboratories, as well as within laboratories, was strictly controlled and compartmentalized. Compton and Lawrence favored prior demonstration, but their information about the mechanics of the bomb, particularly those that would affect the possibility of a demonstration, was incomplete. Lawrence held out longest for prior demonstration, but in the end the panel presented a unanimous recommendation for use without prior warning.

I owed Szilard an answer, but felt it inappropriate to mention my talk with Oppie, as I did not feel that he had authorized me in any way to repeat his opinions. Correspondence at Los Alamos was censored, and I believed it highly likely that Oppie would see my letter. I therefore sent him a copy of my letter to Szilard with a handwritten note:

Dear Oppie,

You may have guessed that one of the men "near Pa Franck" whom I have seen in Chicago was Szilard. His moral objections to what we are doing are in my opinion honest. After what he told me I should feel better if I could explain to him my point of view. This I am doing in the enclosed letter. What I say is, I believe, in agreement with your views. At least in the main points. I hope you will find it correct to send my letter to Szilard.

Edward

July 2, 1945

Dr. Leo Szilard
P.O. Box 5207
Chicago 80, Illinois

Dear Szilard:

Since our discussion I have spent some time thinking about your objections to an immediate military use of the weapon we may produce. I decided to do nothing. I should like to tell you my reasons.

First of all let me say that I have no hope of clearing my conscience. The things we are working on are so terrible that no amount of protesting or fiddling with politics will save our souls.

This much is true: I have not worked on the project for a very selfish reason and I have gotten much more trouble than pleasure out of it. I worked because the problems interested me and I should have felt it a great restraint not to go ahead. I cannot claim that I simply worked to do my duty. A sense of duty could keep me out of such work. It could not get me into the present kind of activity against my inclinations. If you should succeed in convincing me that your moral objections are valid, I should quit working. I hardly think that I should start protesting.

But I am not really convinced of your objections. I do not feel that there is any chance to outlaw any one weapon. If we have a slim chance of survival, it lies in the possibility to get rid of wars. The more decisive a weapon is the more surely it will be used in any real conflict and no agreements will help.

Our only hope is in getting the facts of our results before the people. This might help to convince everybody that the next war would be fatal. For this purpose actual combat use might even be the best thing.

And this brings me to the main point. The accident that we worked out this dreadful thing should not give us the responsibility of having a voice in how it is to be used. This responsibility must in the end be shifted to the people as a whole and that can be done only by making the facts known. This is the only cause for which I feel entitled to doing something: the necessity of lifting the secrecy at least as far as the broad issues of our work are concerned. My understanding is that this will be done as soon as the military situation permits it.

All this may seem to you quite wrong. I should be glad if you showed this letter to Eugene and to Franck who seem to agree with you rather than with me. I should like to have the advice of all of you whether you think it is a crime to continue to work. But I feel that I should do the wrong thing if I tried to say how to tie the little toe of the ghost to the bottle from which we just helped it to escape.

With best regards.

Yours,
E. Teller

I had several reasons for wanting to avoid any further controversy on the issue; as an immigrant, I was particularly aware of my political ignorance; I had not taken sufficient time to think through or discuss the future implications of use versus nonuse; and I sincerely wanted to be on friendly terms with Oppie. I have long regretted the fact that I allowed myself to be so easily persuaded.

Immediately after the bomb was dropped on Hiroshima, the feeling of jubilation among many people in Los Alamos as well as Oppenheimer's dramatic quote from the Bhagavad-Gita, "I am become death, the destroyer of worlds," made me most uncomfortable. I eventually felt strongly that action without prior warning or demonstration was a mistake.

I also came to the conclusion that although the opinions of scientists on political matters should not be given special weight, neither should scientists stay out of public debates just because they are scientists. In fact, when political decisions involve scientific and technical matters, they have an obligation to speak out. I failed my first test in Los Alamos, but subsequently I have stood by that conviction.

It is a remarkable coincidence that with few exceptions (Leo Szilard is the most outstanding), those who favored a prior warning to Japan later argued for continued development of weapons, while those who recommended immediate use of the atomic bomb argued after the war for cessation of all further development.

Lewis Strauss, a Washington-based naval officer during the war, knew about the bomb and personally suggested to Secretary of the Navy James Forrestal that the bomb be demonstrated over a forest after warning the inhabitants to evacuate.[3] In his memoirs, he devotes a whole chapter to the last days of the war and calls it "A Thousand Years of Regret." Years later he became the strongest single supporter of a program to develop the hydrogen bomb.

The BBC film correctly indicates the sharp contrast between Oppenheimer's enthusiastic leadership of the laboratory prior to the bombing and his distress following the bomb's actual use. In the early fall of 1945, Oppenheimer passed me on the way to the laboratory. "Touch me," he said. "I just resigned as director." Quite a few of us knew that Oppenheimer was eager to return to the study of physics and that he was talking about "giving Los Alamos back to the Indians." The future of the laboratory was very much in question.

A few weeks later the decision was made to continue the laboratory at Los Alamos, and when Norris Bradbury took over as the new director, he asked me to stay on as head of physics research. I answered that I would stay under one of two conditions: if we were to have a vigorous program for refining fission weapons, which included at least twelve tests a year, or if we were to concentrate on the hydrogen bomb. In other words, I was fully willing to participate if our work could make a real contribution to the nation's continued military strength. I was not willing to march in place.

Bradbury explained that he wished he could promise to fulfill either of my conditions, but taking political realities into account, he could not do so. I thereupon decided to return to Chicago and work on physics with Fermi and my other friends.

That same evening, Oppie and I were at a party at Deke Parsons's house.[4] Chatting with Oppie, I repeated my afternoon exchange with Bradbury almost verbatim. Oppie said, "And don't you feel better now?" I said, "No." I also remember that on the same occasion Oppenheimer said: "Our accomplishments in Los Alamos have been remarkable, and it will be a long time before anyone can improve on them." I felt less optimistic and could not agree with Oppie's attitude.

Before leaving the war years, I should correct a minor historical inaccuracy in the BBC production. Introducing Oppenheimer's opposition to the hydrogen bomb at the Berkeley summer conference in 1942 enables the producers to suggest future developments but results in a skewed perspective. The hydrogen bomb was the main topic at that conference, and, contrary to the television portrayal, there was no difference of opinion about the propriety of discussing the subject.

Oppenheimer, I was told, actually used the hydrogen bomb in a conversation with Arthur Compton to point out the surprises waiting in the nuclear field and the consequent necessity of establishing a separate laboratory at Los Alamos. The need to pursue such research in the long run was not called into question until after the end of World War II.

Because the United States held a clear monopoly on the atomic bomb in 1945, Oppenheimer began working on and for a plan which the television drama slights. With David Lilienthal and Dean Acheson, he drew up a proposal to place all information about control of atomic weapons in the hands of an international agency. Bernard Baruch presented the plan to the newly created United Nations. The Soviet delegation insisted that before any discussion of how to ensure compliance with the plan could begin, the United States must destroy its nuclear weapons. Since the Soviets were clearly unwilling to come to any reasonable agreement on inspection, the Baruch plan was ultimately dropped.

Today the failure is easy to understand. What we thought we were offering—the secrets of atomic explosives—the Soviets had already gained through their very efficient spy system.

In 1949 I returned to Los Alamos on a full-time basis. The political climate had not improved. Few people seemed to share my concern about the possible progress in development of nuclear weapons in the Soviet Union. However, I had decided to make whatever contribution I could to our own defense.

In September of that year, while in England, I visited with Sir James Chadwick, who had been the leader of the British delegation to Los Alamos. I made an unflattering comment about General Leslie Groves (who headed the whole atomic energy project), and Chadwick, ordinarily a reticent man, became remarkably effusive. He told me that I did not properly appreciate General Groves's dedication and efficiency. Without Groves, Chadwick insisted, the Manhattan project would never have been successful. American scientists had no sense of what it meant to have one's home and family endangered. Their determination and dedication were apt to be too little and too late. After dinner, he walked me back to my hotel. His last words carried unusual emphasis: "Remember what I told you. You may need it."

A few hours after I was back in the United States, it dawned on me that during our conversation Chadwick probably had already known what I had just learned: that the Soviets had exploded an atomic bomb. (An interesting footnote to this event is the fact that without the detection system that was introduced shortly before at the insistence of Lewis Strauss, the United States might have remained in ignorance of the Soviet bomb.) It was then that I called Oppie and was advised, as the film described, that I should "keep my shirt on." It was not the first time since the war had ended that Oppie had made it clear that he was uninterested in using his great talents on defense research problems again.

The BBC production contains to my mind one large flaw. It concerns the position of Lewis Strauss, who at the time of the Oppenheimer hearings was the Chairman of the Atomic Energy Commission (AEC). Strauss appears in the film as one of Oppenheimer's main antagonists.[5]

Early in December 1953 I went to Strauss's office for a prearranged meeting on some laboratory-related matters. He had been unexpectedly called away, so I waited. He returned in uncharacteristic agitation and led me immediately into his office. Pledging me to discuss the issue no further, he told me of the reason for his late arrival and his distress.

I kept Strauss's confidence for many years, but any obligation for silence lapsed long ago. Strauss was appalled because President

Eisenhower had called him to the White House and told him to institute official proceedings to review Oppenheimer's security clearance. Strauss told me with real fervor of his hope that the President's decision would be reversed or at least modified. He foresaw disastrous consequences should Oppenheimer's clearance be called into question.

My experience leaves no room to doubt that Lewis Strauss, far from bringing about those proceedings, wanted to prevent them. Whether Strauss merely foresaw difficult times for the Atomic Energy Commission or whether he had an insight into the future effects on the scientific community, I have no way of knowing.

The film's mistaken sequence—where removal of classified material from Oppenheimer's home occurs before Oppenheimer knows that he has lost his right to retain classified material—and the portrayal of an imaginary meeting of Strauss and Nichols to plot against Oppenheimer create a misleading picture of Strauss. In reality, Lewis Strauss was a sensitive man with a most demanding code of honor. While he knew of the impending troubles, he did not disturb Oppie during his European vacation. As soon as Oppie returned, Strauss called him in to discuss the problem. Strauss explained that a former high-ranking official had written a letter to the President accusing Oppenheimer of disloyalty, that Oppie had the choice of resigning or having a hearing, and that his clearance would be temporarily suspended either way.[6]

Oppenheimer asked how long he had for his decision on resignation or hearing, and Strauss explained that because he had already delayed some weeks, he would appreciate a decision the next day. The classified papers were picked up after that interview. Strauss had not warned of that specifically, but Oppenheimer had had years of experience with security practices and should have not been taken by surprise.

Another detail in that section of the film is in error. When I was called to testify at the hearing, I was, as is shown, met by the attorney for the Atomic Energy Commission, Roger Robb. However, Robb did not give me the FBI file on Oppenheimer. Instead, Robb asked me how I would testify—for or against Oppenheimer's clearance. I had no difficulty with my reply: I would testify for Oppie's clearance. Robb then said he wanted to read a part of the testimony to me. I was a little uncomfortable about it, but an earlier incident seemed to me to have a bearing on what was now appropriate.

Early in 1954, when the question of Oppenheimer's clearance

had become public knowledge, I had met Oppie at a small scientific meeting. I expressed my regrets about his problem. He asked me whether I believed he had behaved in a "sinister" manner. I said that I certainly did not. He then asked me as a favor to go and talk with his lawyer. I agreed to do so. Oppenheimer was not present at the interview, and his lawyer told me nothing I didn't already know.

Having been briefed by Oppenheimer's lawyer, I could find no grounds to refuse Robb. Robb then read Oppenheimer's sworn testimony concerning the Chevalier affair. As the film suggests, that issue proved to be the turning point of the hearing. Oppenheimer testified that he had voluntarily gone to Army security officers with a distorted story (which in the end ruined the life of a friend, Haakon Chevalier). He had told the intelligence officers that Chevalier had asked three scientists to provide information to the Soviets about the atomic bomb project.[7] When asked why he had offered that misleading story, Oppie replied, "Because I was an idiot." I will never forget the shock that this portion of the testimony produced in me. Robb asked me again, "Should Oppenheimer be cleared?" I could only tell him I did not know.

My reluctant testimony, given minutes later, was that I definitely considered Oppenheimer loyal, but because his actions appeared confused and complicated, I would personally feel more secure if public matters would rest in other hands. I was convinced then and continue to believe now that the hearing should never have occurred.

The historical importance of the Los Alamos years are comparatively easy to grasp because of their clearly visible consequence—the use of an incredibly powerful weapon and the end of a terrible war. However, the consequences of Oppenheimer's security clearance are difficult to discern outside the scientific community. They are hardly hinted at in the television drama. Oppenheimer's loss of security clearance partly introduced and partly solidified a deep division in the ranks of American scientists.

After the two events—the use of the atomic bomb and Oppenheimer's loss of clearance—the great majority of scientists felt that it was wrong to work on new weapons. A minority of scientists, to which I belonged, believed it imperative to work on such weapons if the United States was to be able to defend itself and the free world. According to the perception of that minority, the events of the past thirty-five years have demonstrated that while the danger from a ruthless adventurer, Hitler, was more immediate, the danger from the patient, unrelenting leaders in the Kremlin is in reality greater.

Many scientists have never forgiven the damage done to Oppenheimer's reputation. While the origin of the feeling of distrust may have vanished from memory, the residual effect in the scientific community remains. The Oppenheimer hearing was truly as tragic as Strauss feared and, combined with the bombing in Japan, has caused some people today to cry, "A plague on both your houses!" To consider the two superpowers equally at fault seems about as justified as evaluating one's own bad case of acne as equal in seriousness to a neighbor's case of bubonic plague.

One incident depicted in the film is true in spirit but lacks any factual basis. I could very honestly have said on many occasions to Oppie, "I wish I understood you better." Of course, I never said that. But having read Haakon Chevalier's books about Oppenheimer, I have wished for understanding even more intensely.[8] Those books give evidence that Oppie's early leftist associations should not be used to support the conclusion that he was a dangerous Soviet sympathizer. At the same time, the books provide a hint of the unknown depths that were Oppenheimer's personality. I remain totally unable to form an opinion of what his motives were.

The BBC film does not reveal the truth, nor does it provide an explanation. But it gives a glimpse into some of the causes of the confusions and divisions from which people in the free world suffer. We continue to need insight into the contradictions and painful events surrounding that remarkable, ingenious, and paradoxical person, J. Robert Oppenheimer.

8

The Hydrogen Bomb:
The Work of Many People

Few modern technical achievements spring, full-grown, from the mind of a single individual. Success demands teamwork. It depends upon dozens of ideas and hundreds of skills involved with conception and theory, extensive calculations, and finally the actual engineering and construction of the device. Successful development of the hydrogen bomb in the United States was based on that kind of teamwork.

Another story, quite different and quite false, often is presented. A public uninterested in the enormous complexities of modern scientific-technical developments frequently is satisfied with outrageous oversimplifications. Too often only the name of a single individual is mentioned. People are left with the impression that he alone was responsible. Such a representation is both untrue and unjust. An emphasis on the interaction of many different minds and the contribution of many ideas would come closer to the truth and to the real excitement of modern accomplishments.

Inspiration for a hydrogen bomb came from the sun and the stars. A native of Russia, George Gamow, initiated the theoretical work in the United States that ultimately led to the biggest man-made explosion.

Gamow escaped from Soviet Russia in 1933. The next year he joined the physics department of George Washington University, and at his suggestion I joined him there a year later. He infected me with his curiosity about what keeps the sun shining, and he inspired the first pure research into stellar energy in this country. Our early studies were abstract, with no thought of practical application.

Six years before coming to George Washington University, Gamow had reported in the Soviet Academy of Sciences on the work of Robert Atkinson of Britain and Fritz Houtermans of Germany suggesting that the apparently inexhaustible energy of the stars was released by the collisions of atomic nuclei. Those tiny particles, minute even compared with atoms, contain a million times more energy than that released in a chemical reaction. That tremendous energy, stored in the nuclei for billions of years, is released only when the nuclei collide; and collision is usually prevented by the electrical repulsion of the nuclei. Exceedingly high temperatures deep in the interiors of stars, Gamow reported, set up a thermal agitation permitting an occasional collision of nuclei. That leads to a coalescence or fusion of small nuclei into larger units. (The process is the very opposite of fission, which at that time was still undiscovered.) The energy so released produces the brilliance of the stars and the heat radiated by the sun.

When Gamow finished his lecture, he was approached by a high-ranking Soviet official, Nikolai I. Bukharin, ousted by Stalin from a preeminent position and destined eventually for execution, who then was assigned to monitor scientific developments. Obviously excited, Bukharin asked Gamow if the nuclear processes of the stars could somehow be simulated by man for a direct application on earth. Bukharin offered to let Gamow use the Electric Works of Leningrad for a few hours each night to experiment with the possibilities of creating thermonuclear energies. Gamow declined the offer. He said that thermonuclear reactions—the interactions of atomic nuclei at high temperatures—could not be induced at the earth's relatively low temperatures.

At George Washington University in the late 1930s thermonuclear problems became for us a kind of game, an intellectual exercise. We decided that the best candidate for any thermonuclear reaction was the lightest of the elements, hydrogen. This element seemed most abundant in the stars and the sun, and we knew hydrogen nuclei could approach each other most easily, because they carry the lowest electrical charge and repel each other least strongly among all nuclei. In the spring of 1938 Gamow called a conference to consider thermonuclear problems in detail. We accomplished little at the conference except to pose the questions with some clarity. The answers came within a few months. Gamow, with Charles Critchfield and Hans Bethe, succeeded in determining what reactions keep the stars going. They also managed to reconstruct the stars' development, changes in appearance, and the final exhaustion of their energy. Bethe's work was most remarkable: He made a systematic study of every conceivable thermonuclear reaction, catalogued all the meager experimental data of the time, and made some marvelously enlightened guesses about nuclear reactions that had not yet been proved in experiments. His treatment was so complete that nothing useful could be added to his work during the next decade. Bethe proved himself the champion at Gamow's game.

But the research was, in every respect, a game. None of us expected to be able to duplicate the conditions found in the interior of stars, conditions that we knew would be necessary for a thermonuclear reaction. We knew of nothing on earth that could deliver the concentrated energy and heat necessary to fuse nuclei. Then, in December 1938, Otto Hahn and Fritz Strassman discovered fission. Albert Einstein wrote his famous letter to President Roosevelt the next year, and work in the United States started toward the grim reality of the atomic bomb.

I moved to Columbia University in 1941 and began devoting my full energies to the atomic bomb project. But thermonuclear questions were not forgotten. Some imaginative scientists, years before a workable atomic bomb was developed, began wondering whether the concentrated energy of a fission explosion could become the trigger for a thermonuclear bomb. In early 1942 I worked with Enrico Fermi on fission problems at Columbia University, and we usually lunched together at the Faculty Club. Walking back to the laboratory after lunch one day, Fermi posed the question: "Now that we have a good prospect of developing an atomic bomb, couldn't such an explosion be used to start something similar to the reactions in the sun?"

The problem interested me, and during the next few weeks I studied the question. Nuclei of deuterium, or heavy hydrogen, react with each other much more easily than the nuclei of light hydrogen. And Harold Urey had devised a way to separate deuterium from the much more abundant light hydrogen. The process of separation was not too expensive. Substitution of deuterium for light hydrogen, we thought, would represent a long step toward the realization of a comparatively inexpensive thermonuclear reaction. But after a few weeks of concentrated thought I decided that deuterium could not be ignited by atomic bombs. I told Fermi why I thought it could not work and tried to forget all about the intriguing possibilities.

After my negative report to Fermi, my attention was demanded by some details connected with the perfection of nuclear reactors necessary for the production of atomic bombs. Most of the reactor work was being done at that time at the Metallurgical Laboratory in Chicago, and I made plans to move there from Columbia University. At the last minute the plans were changed. Arthur Compton, the prominent and energetic physicist who led the Chicago effort, explained tactfully that I was not needed in Chicago, because all theoretical problems connected with nuclear reactors had been solved. Compton was too kind to tell me the real reason why I could not participate in the top-secret reactor effort in Chicago: I had relatives in Nazi-occupied Hungary, so I could not be cleared for secret work.

About the same time, in the spring of 1942, J. Robert Oppenheimer invited a small group of theoretical physicists to Berkeley, California, for a summer's study of the problems connected with the actual explosion of an atomic weapon. When the invitations were prepared, it became obvious that the United States already was beginning to suffer a shortage of qualified, cleared physicists. Oppenheimer asked that I be given a clearance regardless of my family connections in Hungary, and his request was granted.

As soon as I was cleared, I was invited to both Chicago and Berkeley. I accepted the Chicago invitation but agreed to visit Berkeley and do what I could to further the work there.

At the Metallurgical Laboratory in Chicago, I was assigned to work with another physicist, Emil Konopinski. We were newcomers in the bustling laboratory, and for a few days we had no specific assignments. Both Konopinski and I had been invited by Oppenheimer to the summer session in Berkeley, and I decided that our best contribution to that study might be a detailed review of the reasons why deuterium could not be ignited by an atomic bomb. Konopinski agreed, and we tackled the job of writing a report to show, once and for all, that it could not be done. We wanted no one else to waste valuable time investigating Fermi's curbside suggestion. But the more we worked on our report, the more obvious it became that the roadblocks I had erected for Fermi's idea were not so high after all. We hurdled them one by one and concluded that heavy hydrogen actually could be ignited by an atomic bomb to produce an explosion of tremendous magnitude. By the time we were on our way to California, about the first of July, we even thought we knew precisely how to do it.

In Berkeley, Konopinski and I joined Oppenheimer's group just as it was being formed. The group included J. H. Van Vleck, Felix Bloch, Stanley Frankel, Hans Bethe, and Robert Serber. Although we were called together to investigate the properties and behavior of atomic bombs, all of us were soon engaged in the distant but absorbing question of whether deuterium could be exploded. I presented a rough proof of what could be done and how. My theories were strongly criticized by others in the group, but with the new difficulties new solutions emerged. The discussions became fascinating and intense. Facts were questioned, and the questions were answered by still more facts. As our discussions became more and more detailed, the prospects of success changed almost daily. One day the job would look hopeless; the next day someone would have a bright idea that made everything seem easy. But another member of the group invariably asked a question spotlighting some consideration that had been left out, and the explosion of heavy hydrogen again would appear impossible. A spirit of spontaneity, adventure, and surprise prevailed during those weeks in Berkeley, and each member of the group helped move the discussions toward a positive conclusion. The contributions of Konopinski and Bethe were especially remarkable. Konopinski suggested that, in addition to deuterium, we should investigate the reactions of the heaviest form of hydrogen, tritium. At the time he was only making a conversational guess. It turned out to be an inspired guess. Bethe subjected all of our ideas to the same kind of exhaustive scrutiny he had used earlier to clarify and systematize our knowledge about thermonuclear reactions in the stars.

We were all convinced, by summer's end, that we could accomplish a thermonuclear explosion—and that it would not be too difficult. Oppenheimer was as interested by the prospect as any of us. He concluded: "Now we really need an additional laboratory."

So it was, when the Los Alamos Laboratory was established under Oppenheimer's direction the following year, that exploration of thermonuclear problems was one of the laboratory's objectives. One of the first projects undertaken at Los Alamos was to measure the properties of tritium, a measurement necessary for the thermonuclear work. One of the first buildings constructed at Los Alamos was designed to handle thermonuclear materials. Several of the gifted scientists recruited to work at the Los Alamos Laboratory signed on only because they were intrigued by the thermonuclear possibilities.

The thermonuclear objectives of Los Alamos, however, were sidetracked during the laboratory's first year for two compelling reasons: Successful construction of an atomic bomb proved to be more difficult than anyone had expected, and it became obvious to me that our thermonuclear discussions of the summer before had been incomplete—so incomplete that the new theoretical questions I raised seemed unanswerable, and realization of a thermonuclear explosion seemed most doubtful. The Los Alamos Laboratory, justifiably, gave the highest priority to the field with the greatest promise of early success. Nearly all of the laboratory's theoretical physicists turned their full attention to the atomic bomb project. No matter how difficult it might be, we knew we had to produce an atomic bomb before our enemies could. Work on thermonuclear reactions was all but suspended.

Despite the urgency of the situation, Oppenheimer during those years of struggle with atomic questions did not lose sight of the more distant possibilities. He urged me to continue exploring the thermonuclear field, even though it was beyond the immediate aim of the laboratory. That was not easy advice for him to give or for me to take. It is hard to work apart from others in a scientific community, especially when most people are working toward a goal of the highest interest and urgency. Oppenheimer, Fermi, and many of the most prominent men in the laboratory, however, continued to say that the work at Los Alamos would not be complete as long as the feasibility of a thermonuclear bomb remained in doubt. Until atomic success was verified at Alamogordo on July 16, 1945, however, the thermonuclear program was eclipsed by our country's vital need for an atomic bomb.

After Alamogordo, some of the best scientific minds in the laboratory were applied to thermonuclear problems. Fermi and Bethe were among those who associated themselves with the thermonuclear effort that had been dormant for so long. But their association ended in a few short weeks, before anything could be accomplished. Hiroshima, coming only three weeks after the Alamogordo test, filled many scientists with a moral repugnance for weapons work. Fermi, Bethe, and dozens of others left Los Alamos. Even Oppenheimer, who had supported and urged the thermonuclear effort for years, turned his back on the project. Publicly he announced: "The physicists have known sin." Privately, on the day of Hiroshima, he came to my Los Alamos office for a long talk. He told me that we would not develop a hydrogen bomb. Before Nagasaki, before the war was over, Oppenheimer made it clear to me that he would have nothing further to do with thermonuclear work.

Some members of the small wartime group that had worked on the thermonuclear project at Los Alamos resisted the great exodus of physicists from the laboratory and remained to prepare a summary review of the possibilities of the hydrogen bomb. Stanley Frankel and Nicholas Metropolis worked hardest and longest on the report. They considered the findings we had made in Berkeley in 1942 along with all other relevant data; early measurements made by John Manley, Elisabeth Graves, Marshall Holloway, and Charles Baker; contributions from Fermi and John von Neumann; and the important work of Konopinski, who, with Cloyd Marvin, Jr., proved that a thermonuclear reaction—if initiated on the earth—could under no circumstances spread to ignite the atmosphere or the oceans. The report by Frankel and Metropolis delivered a verdict on the feasibility of a thermonuclear bomb: difficult, but with hard work and concentrated effort, hopeful.

Neither the hard work nor the concentrated effort was in sight. There was no backing for the thermonuclear work. No one was interested in developing a thermonuclear bomb. No one cared. Even keeping Los Alamos alive was an uphill fight, a crucial battle won by the new director of the laboratory, Norris Bradbury. With the existence of the laboratory itself endangered, all-out support for the development of a weapon as devastating as a thermonuclear bomb could hardly be expected. The exceedingly small group of experts whose thermonuclear skills had been developed during the war disbanded. I too left Los Alamos, and not a single member of the wartime thermonuclear group continued to devote his full time and energy to advanced weapons.

But the idea of a superbomb did not die. A very small Los Alamos group headed by Robert Richtmyer kept the spark alive. From my base at the University of Chicago, I traveled to Los Alamos frequently during the years after Hiroshima to confer with Richtmyer's group. From the beginning, our thermonuclear work assumed a new direction and acquired a new style.

I am convinced that if, after Hiroshima, men of Oppenheimer's stature had lent their moral support—not their active participation but only their moral support—to the thermonuclear effort, the United States would have shaved four years from the time it took this country to develop a superbomb. But the thermonuclear work received almost no support in the last months of 1945 or in 1946, 1947, or 1948. Many physicists and government officials were convinced that in the atomic bomb we had the weapon ideally suited for our policy. The people were comforted by published pronouncements that the Soviet Union could not achieve an atomic explosion for at least twenty years. Some leaders felt that work on advanced weapons would make the United States appear to be a warmongering nation bent upon an arms race. Then, in the fall of 1949, the first Soviet atomic explosion made us realize that an arms race was no longer a possibility to be avoided but a frightening reality to be faced.

At Los Alamos, the feeling was widespread that this was the time to pursue development of the hydrogen bomb. I felt that the Soviets would follow their development of a fission bomb with a success in fusion. In that case, the Soviet Union would be far ahead of the United States in the field of nuclear weapons. When Los Alamos was established in 1943, it had been understood that thermonuclear possibilities were to be thoroughly explored. After the first Soviet atomic explosion, most of us at Los Alamos felt the time finally had arrived.

Our enthusiasm was not shared by the powerful General Advisory Committee (GAC) of the Atomic Energy Commission, headed by Oppenheimer. That committee often had a determining voice in AEC policies. On October 29, 1949, a month after President Truman's announcement that the Soviet Union had achieved an atomic explosion, the General Advisory Committee met in Washington to give the AEC an opinion on the advisability of undertaking development of a thermonuclear bomb. Committee members, after a roundtable discussion of the problem, voted unanimously against any H-bomb program. The unanimous report included this statement:

> We all hope that by one means or another, the development of these weapons can be avoided. We are all reluctant to see the United States take the initiative in precipitating this development. We are all agreed that it would be wrong at the present moment to commit ourselves to an all-out effort towards its development.

Although the controlling recommendation was unanimous, the GAC report carried a supplementary statement signed by Oppenheimer, James B. Conant, Lee DuBridge, Hartley Rowe, Cyril Smith, and Oliver E. Buckley. In its final paragraph, the majority report said:

> In determining not to proceed to develop the Super bomb, we see a unique opportunity of providing by example some limitations on the totality of war and thus eliminating the fear and arousing the hopes of mankind.

The negative recommendation of the General Advisory Committee was not communicated immediately to Los Alamos. An effort was made to keep congressional leaders from knowing that scientists close to the problem might disagree with the GAC report.

A few days after the GAC meeting, I was on my way from Los Alamos to Washington to keep an appointment with Senator Brien McMahon, chairman of the Joint Congressional Committee for Atomic Energy. I stopped to see Fermi in Chicago. Despite our very close personal relationship and his knowledge of my almost desperate interest in the thermonuclear effort, he insisted that he could not even give me an indication of the GAC decision. But it was clear from the tenor of his remarks that certainly Fermi and possibly the entire GAC did not favor an all-out crash program. While I was in Fermi's office, I received a telephone call from John Manley, secretary of the General Advisory Committee, who also was associate director of Los Alamos. Manley asked me to see him on my arrival in Washington.

He met me at the railroad station and suggested that I not go to see Senator McMahon. I asked why not. He replied that it would be unfortunate if congressional leaders thought scientists had a divided opinion on the thermonuclear question. I told Manley that I had an appointment with Senator McMahon and intended to see him. Manley insisted that I should not. I offered to telephone Senator McMahon and tell him that I was canceling my visit because I had been asked not to see him. Then Manley gave up, saying: "All right. You better go and see him."

I still did not know the contents of the GAC report when I saw Senator McMahon, and he did not reveal them to me. He did, however, use strong words in reference to the report even before I had an opportunity to ask about it. He said: "I read this report, and it just makes me sick." Still a little mystified about the actual recommendation of the GAC, I told Senator McMahon that I considered it vital to the nation's defense that we proceed with the thermonuclear work. He assured me that he would do everything in his power to make the thermonuclear bomb a reality.

Almost two weeks passed before I had certain knowledge of the GAC recommendation. Manley, back in Los Alamos, asked me into his office and showed me both the minority and majority reports. I could see little difference between them, and I was certain that the thermonuclear effort had been effectively killed.

I was mistaken. The GAC report did stop work on the thermonuclear bomb, because it was tantamount to an explicit instruction to that effect. After a few days, however, the implications of the report began to sink in. It seemed to restrict the Los Alamos scientists to minor improvements in the old field of fission. But many of the scientists, especially the younger men, found it difficult to control an adventurous spirit urging them to get into the newer field of thermonuclear reactions. The GAC report seemed to state the conflict rather bluntly: As long as you people work very hard and diligently to make a better atomic bomb, you are doing a fine job; but if you succeed in making real progress toward another kind of nuclear explosion, you are doing something immoral. The scientists reacted psychologically to that. They got mad. And their attention was turned toward the thermonuclear bomb, not away from it.

The psychological reaction to the GAC report, the scientific anger, certainly could not have produced a hydrogen bomb all by itself. Solution of the theoretical and engineering problems involved in the thermonuclear program required an intensive effort, a concerted action impossible to achieve in a laboratory instructed not to work on the problem. Without a clear go-ahead, Los Alamos could not have produced a hydrogen bomb. Mere anger was not enough. A decision was needed. President Truman was urged to make that decision by AEC Commissioner Lewis Strauss, Senator McMahon, and other members of the Joint Congressional Committee for Atomic Energy.

Ironically, the man who gave our atomic secrets to the Soviet Union also had an important influence on the decision to proceed with the hydrogen bomb. Klaus Fuchs, who was at Los Alamos when we reviewed all we knew about thermonuclear reactions after Hiroshima, confessed in late January 1950 that he had passed secrets to Communist agents. Four days after Fuchs's confession, President Truman overrode the recommendation of the GAC and directed the Atomic Energy Commission to go ahead with the hydrogen bomb. What role the confession of Fuchs played in his decision, I do not know. It has been reported that Truman asked about the hydrogen bomb, "Could the Russians produce it?" Upon receiving a positive answer, he decided to go ahead.

The presidential directive was not a complete surprise to me. A few days before President Truman's decision was announced, I met Oppenheimer at a conference on atomic energy. He made it clear that a top-level decision was being made, and that it probably would direct development of a hydrogen bomb. Recalling his effective leadership of the laboratory during the war, I asked Oppenheimer whether he really would go to work on the hydrogen bomb if President Truman did authorize an all-out thermonuclear program. His reply was negative.

Although I was prepared for the presidential decision of January

29, 1950, I was not prepared for the language of the decision. President Truman directed the AEC to *continue* its thermonuclear program, giving the impression that we could produce a hydrogen bomb simply by tightening a few more screws. People understood from his announcement that the job was almost done. Actually, work had hardly begun. We had eight years of thermonuclear fantasies, theories, and calculations behind us, but we had established no connection between theory and reality. We needed a thermonuclear test.

I was still associated with the Los Alamos Scientific Laboratory when President Truman announced his decision. But, distressed by the opposition of the GAC, I had accepted an appointment as professor of physics at the University of California, Los Angeles, and planned to begin teaching in the fall of 1950. President Truman's decision changed my plans. I had suggested the thermonuclear approach then being pursued at Los Alamos. Naturally, when our efforts received the presidential go-ahead, I wanted to see the work through to completion and find out whether my ideas were right or wrong. Having argued strongly for an all-out thermonuclear program, I felt I had no choice but to attempt to do a job that appeared as difficult as the development of the atomic bomb itself—and was to be undertaken without the world-renowned physicists who had been involved in the wartime atomic effort. I stayed on at Los Alamos.

Immediately and almost simultaneously we tackled two vital problems. Both were mathematical: a detailed review of the favored design of the hydrogen bomb, and preparations for the world's first thermonuclear test.

I was vitally interested in the review of the favored hydrogen bomb design. That design was based upon one of my ideas. The detailed reconsideration of the most likely H-bomb design was started on two fronts, and it soon began to resemble the classic race between the tortoise and the hare. On one front, instructions and information were prepared for what then was the fastest electronic computer. On the other front, an ingenious mathematician, Stanislaw Ulam, and his hard-working friend Cornelius Everett undertook the same computational task by straightforward hand execution. Mathematical ingenuity and hard work won the race. Ulam's results were available even before completion of the lengthy instructions for the electronic brain.

Ulam's first partial results were disquieting. His more complete answers were most discouraging. I could hardly believe them. Ulam's findings seemed to conflict with earlier machine-made calculations. Ulam's work indicated that we were on the wrong track, that the hydrogen bomb design we thought would work best would not work at all.

I decided to wait, before becoming too upset, for the more detailed and accurate results from the electronic computer. When those results were in a few weeks later, Ulam's calculations were substantiated. It appeared clear that the plans we had considered most hopeful had to be revised. Our theories of thermonuclear principles seemed to be on safe ground; we knew what had to be done. But we could no longer point to a particular device and say: "This is how to do it."

Ulam's calculations about bomb design made it absolutely necessary to confirm the remaining parts of our thermonuclear theories. If we were to proceed toward development of a practical hydrogen bomb, we knew we had to proceed in confidence with ideas that were solid and sure. We had to establish a connection between theory and practice. We needed a significant test.

Bomb design was moved to a low priority while we turned our most intense efforts toward preparations for a test that would establish beyond all doubt that a thermonuclear explosion was possible.

Our first job was to make detailed calculations anticipating the results of the first thermonuclear test. Because of the shortage of high-speed electronic computers, much of that arduous work also had to be done by hand. Under the supervision of John Wheeler, Rolf Landshoff, and Robert Richtmyer, a great amount of numerical data was turned out by the people in the Los Alamos computing division. The data were needed to compare the test results with our theory of thermonuclear burning.

Johnny Wheeler was the one internationally known physicist who joined us when the work on the hydrogen bomb became urgent. He arrived one evening in an exhausted state from France, where he had pursued scientific work during a sabbatical. At supper, I told him of a new laboratory-wide decision to speed up our work by adopting a six-day week. Johnny listened but could barely keep his eyes open. Next morning I saw him again at breakfast, fresh and alert. Before opening his egg, he told me, "Edward, last night before I went to sleep I read one sentence in the Bible: 'Six days shalt thou labor.' "

During the last half of 1950 and the early months of 1951, we built a most complex apparatus to record the results of the test explosion, a blast expected to last for only a small fraction of a second. The observation equipment had to be sensitive and fast. It would be vaporized by the explosion but had to record and transmit the blast's effects before being destroyed. Under the direction of Alvin Graves, Frederick Reines, Jack Clark, and William Ogle, crews from Los Alamos as well as the Naval Research Laboratory and the University of California Radiation Laboratory installed the mass of intricate equipment on the Pacific island of Eniwetok.

Meticulous, systematic Hans Bethe was drafted to review our theoretical calculations and measurements. He okayed our work, and we won AEC approval for the first thermonuclear tests. The first test was included in a series of operations code-named Greenhouse.

During the months between approval of our calculations and the Greenhouse explosion in May 1951, I was able to concentrate on the problems of constructing an actual hydrogen bomb. The contraption that we built for the Greenhouse explosion was not a bomb but a purely experimental device designed to establish a scientific principle.

Carson Mark, chief of the Los Alamos theoretical division, in those months before Greenhouse recited some of the difficulties of practical bomb design for a visiting admiral. Mark, who delighted in his contempt for the military mind, paraphrased the admiral's reply and later told me: "He behaved like any other admiral. His reaction was, 'Damn the torpedoes. Full speed ahead.' "

Carson's story irritated me. Then I began to wonder whether we could find detours around the "torpedoes." The detours eventually led to a new approach.

I was guided by two convictions: Ulam had shown that our original ideas about the construction of a practical bomb were incomplete. I was equally certain in my own mind that if we failed in our first attempt to build a practical hydrogen bomb, the General Advisory Committee would recommend abandonment of the entire thermonuclear program. Initial success was imperative. There would be no second chance.

I approached the problem by attempting to free myself entirely of our original concept. That done, it soon became obvious that the job could be done in other ways. During the urgent computations for Greenhouse, many of the hardworking physicists had participated in offhand discussions about the bomb's final design. Some of their ideas were fantastic. Some were practical. None were fully examined. They had been shoved aside by the vital need to complete the calculations for the test. With the theoretical work on Greenhouse finished, those weapons ideas could be examined in detail. Eager to come to grips with the real problem, our group at Los Alamos devoted its full attention to ways of constructing an actual bomb.

On my birthday, January 15, 1951, plans for the forthcoming test were final. I was ready with a new proposal for eventual practical execution. Bradbury, the director of the laboratory, refused to discuss any future plans.

About February 1, 1951, I suggested my new approach to Frederic de Hoffmann. Acting on my suggestion, he made a fine calculation. The results, showing how a thermonuclear bomb could be constructed, were contained in a report that I thought would be signed jointly. De Hoffmann, however, had other plans. He signed the report with only my name, arguing that the suggestion counted for everything and the execution for nothing. I still feel sorry that I consented.

To some who were not closely connected with the Los Alamos effort, our report may have appeared as unexpected and ingenious. Actually, it was the result of hard work and hard thought by many people. The thoughts were incomplete, but all the fruitful elements were present, and the various ideas and suggestions would undoubtedly have been crystallized within a short time into something concrete and provable. If the Los Alamos laboratory had continued to function after Hiroshima with a full complement of such brilliant people as Oppenheimer, Fermi, and Bethe, I am convinced that someone would have had the same idea much sooner—and we would have had the hydrogen bomb in 1947 rather than in 1952. I am just as certain that if we had not proved the practicality of the hydrogen bomb in our report of March 1951, America's thermonuclear effort would have been junked and the first successful hydrogen bomb would have been completed in the Soviet Union.

During March and April of 1951, I urged the feasibility of constructing a hydrogen bomb upon anyone who would listen. Early in March I discussed the report in detail with Norris Bradbury, the director of the Los Alamos Laboratory; Carson Mark, head of our theoretical division; and others at the laboratory. In April I explained my ideas to Gordon Dean, chairman of the Atomic Energy Commission. Dean seemed interested but somehow distracted. After leaving his office, I discovered the reason for the distraction: The zipper on my trousers had failed, and my fly was open. Dean remembered my open fly, not my ideas. Two months later, during another presentation, he seemed to be hearing the ideas for the first time. But in the meantime he had told a magazine reporter that I was a "brilliant if somewhat disarrayed scientist."

Our proof that a practical hydrogen bomb could be readily constructed was based, of course, on theoretical calculations that had not been verified experimentally. That verification, the basic proof needed before making a real H-bomb, came with Greenhouse. Few scientific experiments have been conducted under conditions as exotic or in a place as beautiful as the setting for the first thermonuclear explosion on an atoll in the Pacific Ocean. Rising early that May morning, we walked through the tropical heat to the beach of Eniwetok's placid lagoon. We put on dark glasses, as had been done for the test in Alamogordo. Again we saw the brilliance of another nuclear explosion. Again we felt the heat of the blast on our faces, but still we did not know if the experiment had been a success. We did not know whether or not the heavy hydrogen had been ignited. We did not know whether we had merely seen the explosion of the triggering atomic bomb or actually had witnessed the world's first man-made thermonuclear reaction. The mushroom cloud we saw rising beyond the lagoon showed only that we had been successful in asking a question. The answer had to come from the reports of the recording instruments.

Time was required to gather and interpret those reports, and the twenty-four hours following the test were filled with anxiety. That afternoon, to break the tension, Ernest Lawrence invited me to swim with him in the lagoon. When I came out of the water to stand on the white sands of the beach, I told Lawrence that I thought the experiment had been a failure. He thought otherwise, and bet me five dollars that we had been successful in igniting heavy hydrogen and producing a thermonuclear reaction.

I was hardly awake the next morning when Louis Rosen, one of the most ingenious experimentalists, burst into my quarters to announce: "I have the evidence! Only one piece, but I have evidence that the test was a success. Please, please tell no one until it is verified." I promised. Lawrence would be leaving the island that morning before additional readings could be made, but I would keep my promise to Rosen. I told no one and waited as long as I could for final verification. When Lawrence left for the air strip, I could wait no longer, so I ran after his jeep and silently handed him five dollars. I knew that success at Greenhouse ensured the successful construction of a hydrogen bomb along the lines detailed in the report to which de Hoffmann had signed my name two months earlier.

A month after Greenhouse, the Atomic Energy Commission called a significant roundtable conference to determine the best way to build a hydrogen bomb. The meeting was held June 19 and 20 at the Institute for Advanced Study at Princeton. Oppenheimer, as chairman of the Weapons Committee of the GAC, presided. Members of both the AEC and the GAC attended, along with Los Alamos scientists.

I was amazed when Carson Mark, in his presentation, did not mention the hydrogen bomb report that I had handed him three months before. My amazement multiplied when Gordon Dean, still chairman of the AEC, spoke without mentioning the same report, which I had explained to him two months earlier. My amazement approached anger as other scientists and officials who knew of the report spoke without referring to it. Finally, I could contain myself no longer. I insisted on being heard. My demand occasioned a spirited debate, but it was decided that I should be allowed to speak. I walked to the blackboard and again went through the theory and calculations that were already familiar to half the men in the room.

Response to the theory, now supported by the experimental evidence of Greenhouse, was enthusiastic and unanimous. Gordon Dean, who apparently had found my sloppy dress an insurmountable distraction two months before, later testified:

> Out of the meeting came something which Edward Teller brought into the meeting within his own head, which was an entirely new way of approaching a nuclear weapon. I would like to be able to describe that but it is one of the most sensitive things we have left in the atomic energy program. (censored) At the end of those two days we were all convinced, everyone in the room, that at last we had something for the first time that looked feasible in the way of an idea. (censored) I remember leaving that meeting impressed with this fact, that everyone around the table without exception, and this included Dr. Oppenheimer, was enthusiastic. . . . . [1]

During the months preceding and following the Princeton meeting, ingenious and reliable calculations were carried out in connection with this new kind of nuclear explosion. Marshall Rosenbluth, Conrad Longmire, Lothar Nordheim, and many others made accurate predictions about details of the way our new device would function.

Under the leadership of Marshall Holloway, a new test was prepared on one of the islets of the Eniwetok chain, Elugelab. On November 1, 1952, that islet was wiped off the face of the earth by the first full-scale thermonuclear explosion.

I was not on hand for the explosion of the first hydrogen bomb. I left Los Alamos exactly one year before that momentous event. The battle for the thermonuclear bomb had been won at the Princeton conference, and I was drawn to the fight for establishment of a second weapons laboratory. On November 1, 1951, I left Los Alamos and took a last look at the gold-lettered poem framed on my office wall, a prayer fondly quoted by Lewis Strauss:

Providence, who watches over children, drunkards and fools
With silent miracles and other esoterica,
Continue to suspend the customary rules
And protect the United States of America.

My inability to see development of the thermonuclear bomb through to a final, successful conclusion was a great disappointment. But there were good reasons for my leaving Los Alamos.

It was an open secret, among scientists and government officials, that I did not agree with Norris Bradbury's administration of the thermonuclear program at Los Alamos. Bradbury and I remained friends, but we differed sharply on the most effective ways to produce a hydrogen bomb at the earliest possible date. We even disagreed on the earliest possible date itself, on the timing of our first hydrogen bomb test. The dissension with Bradbury crystallized in my mind the urgent need for more than one nuclear weapons laboratory.

I knew that science thrives on friendly competition, on the fostering of different points of view, and on the exchange of ideas developed in different surroundings. I knew, too, that a single group of scientists working together can easily become fascinated by special aspects of a development to the neglect of other hopeful approaches. My conviction grew that the safety of our country could not be entrusted to a single nuclear weapons laboratory, even a laboratory as excellent as Los Alamos. That conviction was hardened by a growing awareness, as our work progressed at Los Alamos and our thermonuclear knowledge increased, that we were pioneering a big new field of weapons development. I began to doubt that one laboratory would be physically capable of handling all the work that had to be done. Weighing all of those ideas and circumstances, I came to the inescapable conclusion that at least two weapons laboratories, working in cooperation but also in the traditional American spirit of competition toward the mutual goal of adequate national defense, were vital to the future of the United States.

I also concluded that I could advocate establishment of a second weapons laboratory most effectively if I were not associated with the existing Los Alamos Laboratory. So, regretfully, I left Los Alamos and returned to the University of Chicago.

During the following year, the first hydrogen bomb was developed and perfected at Los Alamos. My work during that year took me from the University of Chicago to the University of California. In October 1952, while in California, I was invited to the South Pacific to watch the explosion of "Mike," the world's first full-scale hydrogen bomb. I very much wanted to see the explosion of the device that had consumed my energies and that had dragged me into so many arguments. But I knew that I really was not needed at Eniwetok. So I compromised with my desire: I attended the first hydrogen bomb explosion by watching the sensitive seismograph at the University of California in Berkeley.

On the morning of November 1, 1952, I was escorted into the dark basement room where the seismograph was writing its tremulous record on a photographic film with a fine beam of light. The spot of light seemed unsteady. It moved more than it should have to record the continuous, minute trembling of the earth or the pounding of ocean waves on our shores. The light's jumpiness was due to the movements of my own eyes; in the darkness my eyes were not steadied by the surrounding picture of solid objects. I braced a pencil and held it close to the luminous point. Now the point seemed steady. The earth was quiet. That was about the time of the shot at Eniwetok. Nothing happened on the seismograph, and nothing could have happened. It took about fifteen minutes for the shock of the explosion to travel, deep under the Pacific Basin, to the California coast.

I waited impatiently and watched the seismograph make a time signal each minute. At last the time signal came that had to be followed by the explosion's shock, and there it seemed to be: The spot of light danced wildly and irregularly.

But I almost convinced myself that what I had seen was the motion of my own hand and the pencil it was holding rather than the signal from the first hydrogen bomb. The film was taken from the seismograph and developed, and the tracing appeared on the photographic plate. It was clear, big, and unmistakable. It had been made by a wave of compression traveling thousands of miles and bringing positive assurance that our first hydrogen bomb had been a success.

I believe that everyone who worked on the hydrogen bomb was appalled by its success and by its possible consequences. I believe most people who were connected with the effort were driven by the realization that the work was necessary for the safety of our country. I also know that our work contributed to science in a variety of ways, a consideration I do not want to forget.

We would be unfaithful to the tradition of Western civilization if we shied away from exploring what man can accomplish, if we failed to increase man's control over nature. The duty of scientists, specifically, is to explore and to explain. That duty led to the invention of the principles that made the hydrogen bomb a practical reality. In the whole development I claim credit in one respect only: I believed, and persisted in believing, in the possibility and the necessity of developing the thermonuclear bomb. My scientific duty demanded exploration of that possibility.

Beyond the scientific responsibility to search the horizons of human knowledge, the responsibilities of scientists cannot be any greater than those of any other citizen in our democratic society. The consequences of scientific discoveries are the responsibility of the people. Every citizen, whether a politician or a farmer, a businessman or a scientist, has to carry his share of the greater responsibility that comes with greater power over nature. But a scientist has done his job as a scientist when that power has been demonstrated.

*　　*　　*

Author's postscript. The first version of this essay, published in *Science* in February 1955, was written in the midst of acrimonious controversy. After writing the article, I did not know if I should publish it. During that period of uncertainty, I received some terrible news: My good friend, the great Enrico Fermi, was dying of cancer. I went to Chicago to see him one last time.

Characteristically, he did not want to talk about himself. He wanted to talk about what I was doing. I told him of this paper and my doubts about publishing it. "Do you have it with you?" he asked. The paper was in the one bag with which I traveled.

Fermi read it slowly and carefully and, when he finished, asked, "Tell me the reasons for your hesitation." I tried to comply. Then Fermi, who made the first suggestion of the possibility of a hydrogen bomb, but who later in the General Advisory Committee opposed the program to build it, gave me his advice. I followed his suggestions.

Enrico was a great scientist. He was also a memorable man.

9

Two Stories

In 1975 I met Aleksandr Solzhenitsyn for the first and only time. Across a language barrier that barely allowed communication, he asked, "Do you know Sakharov?" I told him I did not. Solzhenitsyn replied, "This is terrible world."

Since that time I have read most of Sakharov's published works, and I now feel I understand Solzhenitsyn's comment. Sakharov and I were born thirteen years and about 1,000 miles apart, to families that valued learning, music, and courtesy. We both became theoretical physicists; we both applied our knowledge to weapons development; and both our lives were changed by the major wars of the twentieth century. The similarities between Andrei Sakharov and me do not stop here, but the points of divergence are so numerous that our stories lose all semblance of a parallel development.

Sakharov entered the university in Moscow, his home town, in 1938. By the time Hitler attacked Russia in 1941, Sakharov's exceptional abilities had been recognized. He was deferred from military service to finish his undergraduate studies. A year later he entered defense work, designing quality control devices. In 1945, in a victorious but sorely damaged nation, he began his graduate studies in Moscow under Professor Igor Tamm, an eminent physicist, and completed them three years later. Then his life took a different turn:

> A few months after defending my dissertation for the degree of Candidate of Doctor of Science, roughly equivalent to an American Ph.D., which occurred in the spring of 1948, I was included in a research group working on the problem of a thermonuclear weapon. I had no doubts as to the vital importance of creating a Soviet super-weapon for our country and for the balance of power throughout the world.[1]

He was at that time twenty-seven years old.

When I was twenty-seven years old, I arrived in the United States. I suspect that, even though I had been through many upheavals, I had enjoyed more pleasures and much greater safety than had Sakharov. Between the ages of six and eleven, I lived in a nation at war, but I was close to its violence only for a single day in 1919, when street fighting occurred during a brief communist regime. I have clear memories of hunger during that period. The real tragedy of World War I was that it killed emerging democracies, both in Russia and in Hungary.

At eighteen I left Hungary, where anti-Semitism closed off most opportunities, to study in Germany. There I entered a community of scholars saturated with the discoveries of the early part of the century, concepts that changed mankind's outlook on the universe in a profound way. I joined in the adventure of freely exploring ideas of all sorts. Finding a spiritual home is considerable compensation for the great shock of leaving one's family and country.

I completed my doctoral work under Werner Heisenberg, a brilliant physicist and a wonderful human being. At twenty-three, I became a research associate at Göttingen, enjoying the company of a wide circle of friends from many nations. Two years later, in February 1933, I noticed the first signal of the terror that was descending on Germany. All the advertisements and election posters disappeared from the kiosks. The Nazis were on their way. I resumed my peregrinations, traveling to Denmark, England, and eventually the United States.

Although by then I had left my native land, and even the continent where I had been educated, my new friends in the scientific community, together with several fellow émigrés, provided continuity in my life. The community of scientists is worldwide. I had reason to believe that I could be at home anywhere.

My first years in the United States were a happy period marked by a steady increase in pure knowledge and the opportunity to share it. But in 1939 I became peripherally involved in what later became the Manhattan Project. During the next year, after the invasion of Belgium and Holland, I made a commitment to work on weapons for the defense of freedom. In 1942 I recognized that a fission bomb made the development of a thermonuclear weapon likely. In April 1943 I joined the research group at Los Alamos.

Sakharov's early experience in weapons work was quite different from my own. When I read his account, I could hardly appreciate its ramifications. I knew he had begun work on the hydrogen bomb in 1948 under Stalin. In 1977, he had described Stalin as "one of the

greatest criminals of the sorely tried twentieth century."[2] That in itself would be a terrible burden to bear.

However, it was not his only burden:

> In 1950 our research group became part of a special institute. . . . Until
> the summer of 1953, the chief of the atomic project was Beria, who ruled
> over millions of slave-prisoners. Almost all the construction was done with
> their labor.[3]

I remember our petty and insignificant annoyances at Los Alamos with General Groves, a military commander who lived worlds
apart from the scientific "eggheads." He was rough, ambitious, and
completely dedicated to the Manhattan project. He appreciated the
scientists somewhat more than the scientists appreciated him, a fact
I came to understand with the passage of years.

The Soviet thermonuclear research program—from its initiation
in 1948, through its successful production of a fission weapon in
1949, until shortly before the successful test of a fusion device in
August 1953—was directed by the head of the secret police, Lavrenty
Beria.

But in 1948 Sakharov may not have thought of anything except
the recent terrible war experiences of his country and the prevention
of any recurrence. I doubt that Sakharov knew, as he entered the
program, that the great Russian physicist Peter Kapitsa had been
placed under house arrest for refusing to work on the Soviet atomic
bomb project. Sakharov's mentor, the Nobel laureate Igor Tamm,
went with seeming willingness into the weapons project. Having just
completed his doctoral studies in theoretical physics, it is unlikely
that Sakharov had any real understanding of Soviet politics.

My first active connection with atomic energy was made at the
request of my friend, Leo Szilard. In Washington, New York, Chicago, and finally Los Alamos, I worked on the atomic bomb. It was
a new field of activity for me, but I was engaged in it in the company
of my friends and associates. I had merely started on the road toward
a more difficult, more responsible, and far less satisfying activity
than theoretical physics.

At the end of the war, the fact that the existence of the atomic
bomb greatly increased the likelihood that a thermonuclear weapon
was feasible made me desperately anxious. For five years I talked
with everyone in sight about the urgent need for research. That was
the beginning of a rift between the majority of my friends in the

scientific community and me. It was a greater shock than leaving the country of my birth.

Finally, in 1950, enough support was gathered for President Truman to give the go-ahead for work on the hydrogen bomb. The project in the Soviet Union was already more than a year old.

If I was separated from my scientific colleagues by my beliefs, Sakharov was separated by his location after 1950 at the secret laboratory in Turkestan and by his life of extreme material privilege. He had a high salary, special housing, a chauffeur, and access to restricted consumer goods. Honors were heaped upon him. He even became a member of the prestigious U.S.S.R. Academy of Sciences in the same year as his mentor, Professor Tamm, thereby becoming the youngest full member in the distinguished history of that academy.

For some individuals, however, the demands of the human spirit make all such things of little importance. Sakharov, writing years later, makes it plain that he became aware of the Soviet caste system and of the cynicism of the leaders. He was anything but comfortable.

In 1961 he began a series of actions that changed his life profoundly. Because history is so quickly forgotten, the preceding events deserve a brief review. Atmospheric testing by both the United States and the Soviet Union had continued for the six years that followed the first H-bomb tests. Then, late in 1958, a "gentlemen's agreement" was established between the Soviet Union and the United States to halt all testing. The United States was convinced that any Soviet preparations for a meaningful test would be noticed early enough to allow us to prepare our own. In the summer of 1961 the Soviet and American negotiators were discussing a final treaty banning nuclear tests.

That same summer, Sakharov was called to a meeting between Soviet atomic scientists and First Secretary Khrushchev. Sakharov reports what occurred after they had gathered around the conference table:

> It turned out that we were to prepare for a series of tests that would bolster up the new policy of the USSR on the German question [the Berlin Wall]. I wrote a note to Khrushchev, saying "To resume tests after a three-year moratorium would undermine the talks on banning tests and on disarmament, and would lead to a new round in the armaments race—especially in the sphere of intercontinental missiles and anti-missile defense." I passed it up the line. Khrushchev put the note in his breast pocket and invited all present to dine. At the dinner table he made the off-the-cuff speech that I remember for its frankness and that did not reflect merely his personal

position. He said more or less the following: Sakharov is a good scientist. But leave it to us, who are specialists in this tricky business, to make foreign policy. Only force—only the disorientation of the enemy. We can't say aloud that we are carrying out our policy from a position of strength, but that's the way it must be. I would be a slob, and not Chairman of the Council of Ministers, if I listened to the likes of Sakharov.[4]

What must it have felt like as a forty-year-old scientist—a man who had worked diligently for the good of his homeland—to discover what was behind the policies of his government? The 1961 Soviet test series took the United States totally by surprise. Announced by Khrushchev the day before it began, the series involved ten multimegaton devices, including one explosive of approximately 60 megatons, the largest ever produced. The United States, having assumed that weapons testing had ended, had no plans ready. We conducted a hastily assembled test series the following spring.

Understanding the intensity of Sakharov's feeling about the 1961 Soviet tests, which involved well over 100 megatons of explosives, requires a little knowledge of the nature of thermonuclear weapons. Making a multimegaton explosive requires less sophistication than making a smaller explosive. Large weapons produce an indiscriminate and militarily worthless amount of damage. The size of U.S. weapons steadily decreased after 1954. By 1961 the Soviets had tested several weapons of smaller size but chose to continue detonating the much larger variety, perhaps for the purpose of intimidation.

The text of Sakharov's Nobel Peace Prize speech addressed the consequences of such a policy.

> This cooperation [between the Western states, the socialist nations, and the developing countries] must be based on mutual trust between open societies, or—to put it another way—with an open mind, on the basis of genuine equality and not on the basis of the democratic countries' fear of their totalitarian neighbors. If that were the case, cooperation would merely involve an attempt at ingratiating oneself with a formidable neighbor. But such a policy would merely postpone the evil day, soon to arrive anyway and, then, ten times worse. This is simply another version of Munich.[5]

Sakharov repeated similar advice in a recent open letter to a physicist who supported the U.S. movement to "freeze" nuclear weapons:

> Precisely because an all-out nuclear war means collective suicide, we can imagine that a potential aggressor might count on a lack of resolve on the part of the country under attack to take the step leading to that suicide, i.e., it could count on its victim capitulating for the sake of saving what could be saved. . . . Take the next logical step—while nuclear weapons

exist it is also necessary to have strategic parity in relation to those variants of limited or regional nuclear warfare which a potential enemy could impose. . . . If the probability of [an all-out nuclear war] could be reduced at the cost of another ten or fifteen years of the arms race, then perhaps that price must be paid. . . .[6]

Sakharov has made an incredible sacrifice in order to share his expertise and understanding with the defenders of peace and freedom. Unfortunately, his specific message seems not to receive sufficient notice in the West.

While breach of faith formed the substance of Sakharov's plea in 1961, he states that the motivation for his act lay elsewhere and began four years earlier:

Beginning in 1957 (not without the influence of statements on this subject made throughout the world by such people as Albert Schweitzer, Linus Pauling, and others) I felt myself responsible for the problem of radioactive contamination from nuclear explosions. As is known, the absorption of radioactive products of nuclear explosions by the billions of people inhabiting the earth leads to an increase in the incidence of several diseases and birth defects, of so-called sub-threshold biological effects. . . . Each series of tests of a nuclear weapon . . . involves tens of megatons; i.e., tens of thousands of victims.[7]

While there was a great deal of speculation about possible low-level radiation hazards in the late 1950s and early 1960s, the substantiating evidence of damage—although intensely sought—never appeared. How could a scientist of Sakharov's ability confuse popular exaggeration with fact? How did he come to place so much emphasis on statements made by Schweitzer and Pauling and to make so cursory an investigation himself? How does an honorable and loyal person deal with the demands of his country and of humanity if they conflict? Surely, no person of any experience can believe that this is accomplished in an easy or a purely logical way.

My teacher and friend Werner Heisenberg faced a somewhat similar dilemma during World War II. Heisenberg was the director of the German atomic bomb project under Hitler. From a technical point of view, there could have been no better choice. However, Heisenberg disliked the Nazis. He was appalled at the widespread imprisonments and went out of his way to save its victims whenever possible. One pertinent example was Fritz Houtermans, a talented German physicist with leanings toward communism.

After Hitler came to power, Houtermans went to Russia. The years between 1930 and 1939 were terrible years in the Soviet Union.

Houtermans, during one of Stalin's purges, was sent to prison, but after the 1939 Hitler–Stalin pact he was shipped back to Nazi Germany, headed for a concentration camp. Heisenberg rescued him by stating that he needed Houtermans to work on the atomic energy project. That, surprisingly enough, is just exactly what Houtermans did. At one point he suggested that bombarding uranium with neutrons would produce a new element (now called plutonium), which could serve as an atomic explosive.

Heisenberg was too loyal to his country to accept escape when it was offered. When, in the summer of 1939, he was urged by his many friends and colleagues in democratic nations to do so, he asked, "Do you desert your brother because he stole a silver spoon?" Unfortunately, the crimes of the Nazis turned out to be incomparably greater.

Heisenberg almost succeeded in building a nuclear reactor, but he could not understand the validity of Houtermans's suggestion. How could that have happened to this brilliant physicist? Heisenberg, like Sakharov, cannot be blamed. Finding any solution to the situation in which they were placed required much more than simple logic. It required what may have been impossible.

The means by which Sakharov came to his decision to act are far less important than the fact that he reached his decision. The decision took independent judgment, real intelligence, and a truly extraordinary amount of courage. I sometimes wonder how many parallels there may be in the history of the world for that kind of spiritual evolution. We have plenty of wonderful stories about sudden conversions. A slow, thoughtful, and consistent change of opinion, made in the face of immense personal loss, is a rarity. It is not something that can be supported on a wave of emotion or by enthusiasm. It requires the determination and the character of a man with unusual integrity.

Sakharov and I might have met at the Second Atoms for Peace Conference in 1958. Sakharov is proud that some of his ideas about peaceful fusion energy generation were presented at that time on his behalf. I was there, happy to start the exchange of information on controlled fusion. That was one step the United States took to decrease secrecy about scientific matters. Sakharov has written extensively about third world energy needs and technology's role in offering the people of poor countries a more hopeful future. We might have seen each other at a conference related to the development of the third world. We have never met. As Solzhenitsyn said, this is a terrible world.

I am now more than three-quarters of a century old. Sakharov is an ailing prisoner of one of the most heartless regimes in a period that has demonstrated man's inhumanity to man in new and ingenious fashions. I am afraid we shall never meet. I still hope that such an event might occur. If it does, I am sure we could disagree on many points within the surprisingly wide but nevertheless strict limits set by reason for all thinking men. But I am also sure that I would find great harmony with a man who, lacking all direct experience with living in freedom, can write:

> I am convinced that international trust, mutual understanding, disarmament, and international security are inconceivable without an open society with freedom of information, freedom of conscience, the right to publish, and the right to travel and choose the country in which one wishes to live.[8]

Sakharov has unified the opinion of scientists in every part of the world where opinions can be freely expressed. By publicizing his story and those of his friends, he has clarified current issues in a manner that could have occurred in no other way. Since Galileo, no other scientist has been honored as Sakharov is. In my opinion, Sakharov is even more deserving of the highest respect than the great Italian. Sakharov has suffered far more, and, once having clarified his thoughts, he has never recanted.

III

THE
UNSTABLE
PEACE

10

To Win the Peace

Many people know that the Manhattan Project was initiated by a letter Einstein wrote to President Roosevelt. Few people know that Einstein wrote a second letter to Roosevelt on March 25, 1945. Both letters were written at the initiative of Leo Szilard.

Six months before the first test of an atomic weapon, Szilard had no doubts of its success. His concern lay solely with its effects on the future. In a characteristically detailed manner, Szilard wrote a memorandum about questions related to the use and future effects of the atomic bomb, particularly about the possibility of an arms race. Einstein's second letter was meant to provide the means for presenting that memorandum to President Roosevelt.

This time, Szilard's informal channels of communication included Eleanor Roosevelt. On the basis of Einstein's letter, she secured an appointment for Szilard with the President—on May 8, 1945.

For several days after April 12, when Roosevelt died, Szilard was at a loss for what to do. But soon after he had an appointment at the White House with Truman's appointments secretary, Matt Connelly. Connelly sent Szilard to see James F. Byrnes, at that time a private citizen in Spartanburg, South Carolina. (Byrnes would later be Truman's Secretary of State.) The meeting with Byrnes was not a success. As Szilard described it:

> I began to doubt that there was any way for me to communicate with Byrnes in this matter, and my doubt became certainty when he turned to me and said, ''Well, you come from Hungary—you would not want Russia to stay in Hungary indefinitely.'' I certainly didn't want Russia to stay in Hungary indefinitely, but what Byrnes said offended my sense of proportion. I was concerned at this point that by demonstrating the bomb and using it in the war against Japan, we might start an atomic arms race between America and Russia which might end with the destruction of both countries. I was not disposed at this point to worry about what would happen to Hungary.[1]

Byrnes never forgot Szilard. About 1971, my physician friend Dr. J. R. Maxfield introduced me to Byrnes. His first comment was directed not to me but to Maxfield. "What?" Byrnes asked. "Another Hungarian physicist? Is he like Szilard?"

In the course of our conversation, Byrnes told me about his meeting with Szilard, who was accompanied on that occasion by Harold Urey. Urey, Byrnes said, had been silent, but Szilard had proved absolutely intractable. He continued to insist that the bomb should not be used. At last Byrnes told him that he would forward his suggestion to Oppenheimer, who was heading a group of scientists that were advising the government about the use of the atomic bomb. Then, Byrnes reported, Szilard was finally satisfied.

Yet Szilard must not have had complete confidence in the answer. On his way from Spartanburg to Chicago he stopped in Washington to see Oppenheimer, who was there for a meeting of the advisory scientists (Compton, Lawrence, Fermi, and Oppenheimer). Szilard told Oppenheimer that it would be a very serious mistake to use the bomb against the cities of Japan, but Oppenheimer didn't share his view. According to Szilard, Oppenheimer said:

> "Well, this [the atomic bomb] is a weapon which has no military significance. It will make a big bang—a very big bang—but it is not a weapon which is useful in war." He thought that it would be important, however, to inform the Russians that we had an atomic bomb and that we intended to use it against the cities of Japan, rather than taking them by surprise. This seemed reasonable to me, and I knew that Stimson also shared this view. However, while this was necessary it was not sufficient. "Well," Oppenheimer said, "don't you think that if we tell the Russians what we intend to do and then use the bomb in Japan, the Russians will understand it?" And I remember that I said, "They'll understand it only too well."[2]

Shortly after his return to Chicago, Szilard set about collecting signatures for his petition, and in late June I received a copy, which I, in ignorance of the earlier discussions, took to Oppenheimer:

A Petition to the President of the United States

Discoveries of which the people of the United States are not aware may affect the welfare of this nation in the near future. The liberation of atomic power which has been achieved places atomic bombs in the hands of the Army. It places in your hands, as Commander-in-Chief, the fateful decision whether or not to sanction the use of such bombs in the present phase of the war against Japan.

We, the undersigned scientists, have been working in the the field of atomic power. Until recently we have had to fear that the United States might be attacked by atomic bombs during this war and that her only defense might lie in a counterattack by the same means. Today, with the defeat of Germany, this danger is averted and we feel impelled to say what follows:

The war has to be brought speedily to a successful conclusion and attacks by atomic bombs may very well be an effective method of warfare. We feel, however, that such attacks on Japan could not be justified, at least not unless the terms which will be imposed after the war on Japan were made public in detail and Japan were given an opportunity to surrender.

If such public announcement gave assurance to the Japanese that they could look forward to a life devoted to peaceful pursuits in their homeland and if Japan still refused to surrender our nation might then, in certain circumstances, find itself forced to resort to the use of atomic bombs. Such a step, however, ought not to be made at any time without seriously considering the moral responsibilities which are involved.

The development of atomic power will provide the nations with new means of destruction. The atomic bombs at our disposal represent only the first step in this destruction and there is almost no limit to the destructive power which will become available in the course of their future development. Thus a nation which sets the precedent of using these newly liberated forces of nature for purposes of destruction may have to bear the responsibility of opening the door to an era of devastation on an unimaginable scale.

If after this war a situation is allowed to develop in the world which permits rival powers to be in uncontrolled possession of these new means of destruction, the cities of the United States as well as the cities of other nations will be in continuous danger of sudden annihilation. All the resources of the United States, moral and material, may have to be mobilized to prevent the advent of such a world situation. Its prevention is at present the solemn responsibility of the United States—singled out by virtue of her lead in the field of atomic power.

The added material strength which this lead gives to the United States brings with it the obligation of restraint and if we were to violate this obligation our moral position would be weakened in the eyes of the world and in our own eyes. It would then be more difficult for us to live up to our responsibility of bringing the unloosed forces of destruction under control.

In view of the foregoing, we, the undersigned, respectfully petition: first, that you exercise your power as Commander-in-Chief, to rule that the United States shall not resort to the use of atomic bombs in this war unless the terms which will be imposed upon Japan have been made public in detail and Japan knowing those terms has refused to surrender; second, that in such an event the question whether or not to use atomic bombs be decided by you in the light of the considerations presented in this petition as well as all the other moral responsibilities which are involved.

Presented: July 17, 1945

As far as I have learned, Szilard had no success in circulating the petition at any laboratory outside of the University of Chicago. In the end he was able to collect a few dozen signatures. He sent them off, against his will but in accordance with the wishes of the signators, through regular channels. As Szilard feared, the petition never reached President Truman.

Szilard's proposal never really had much of a chance. Even if the scientific commission appointed by Truman to make recommendations of the use of the bomb had supported Szilard's idea, it is unlikely that in the wartime setting they would have been heard.

Actually, another effort in the same direction advanced much farther than Szilard's. On April 25, 1945, Harold Stimson, the Secretary of War, briefed President Truman on the atomic bomb project. Stimson reported that the first bomb would be tested about mid-July, that the bombs would be available in about four months, and that each bomb could destroy a whole city. Stimson then added his personal opinion that the "development of this weapon had placed a certain moral responsibility upon us which we cannot shirk without very serious responsibility for any disaster to civilization which it would further."[3]

While Stimson also carried the report of the successful test to Potsdam, the subject of not using the bomb was not discussed there. President Truman was deeply preoccupied with the fate of Europe during the coming winter. After receiving the news of the test, he asked the opinion of only one man, Winston Churchill, who also believed in the urgent need to end the war. Without any further hesitation, Truman took what seemed to be the fastest path to ending World War II.

What would have happened had Roosevelt lived a few years longer? His greatest mistake was that he left no political testament. He might have found a way, without loss of life, to end a horrible war (and the horrors of World War II should never be forgotten). He might have succeeded in introducing the atomic age by showing that incredible scientific progress could be used to make peace without bloodshed.

The end of the war left as part of its inheritance a powerful army and ongoing lend-lease to the Soviet Union. What might have happened had the United States presented a realistic plan for the cooperative control of atomic energy at the time of the Japanese surrender? The alternatives for the Soviet Union were different at that time.

Roosevelt planned the United Nations. He would not have readily consented to the highest international authority's becoming a mere debating society. What were his plans? What if a man of Roosevelt's stature had become the Secretary General? Dramatic action might conceivably have prevented the sad conflict that is now known as the cold war.

Imagination can move fast; actual events moved more slowly. The proposal for cooperation on nuclear energy, which should have been offered in August 1945, was not presented until June 1946. By that time, lend-lease had ended, the mighty U.S. Army was disbanded, and spies had given away our secrets. The proposal presented at the United Nations was accompanied by neither a carrot nor a stick. Stalin was immovable.

The actual offer was drafted by Dean Acheson and David Lilienthal, but it was the brainchild of J. Robert Oppenheimer, the scientific consultant to the project. Oppenheimer had not gone along with Szilard's futile attempt to end the war without actually using the bomb. Now, however, his proposal, known variously as the Acheson Report and the Baruch Plan, seemed a possible solution to an increasingly serious problem.

After the exodus from Los Alamos, Eugene Rabinovich, Clyde H. Hutchinson, and I, in Chicago, had discussed the need to inform the public about the facts and about the difficulties presented by atomic energy. That was the beginning of a new magazine. I proposed the picture of a clock at ten minutes to midnight for its cover. I didn't foresee that the magazine would become a partisan publication or that the clock would run backward or forward at the whim of the editors.

The postwar euphoria was soon replaced by the increasing realization that America was no longer and would never again be a country isolated and made secure by its ocean barriers. Although the nuclear age began in 1945, its reality did not dawn on most Americans until we found out, in 1949, about the first Soviet atomic bomb. Even today, this new situation is called unthinkable. In plain English, it is something we prefer not to think about.

The next chapter was originally an article published in two parts in the early issues of *The Bulletin of Atomic Scientists*. It is still of interest to remember the Baruch Report because of its realism. The contrast between that first proposal and all later attempts at agreement is marked. Real agreement has been replaced by wishful thinking, to which we continue to be addicted.

11

A Ray of Hope

International control of atomic energy is a peculiarly difficult problem. If the control fails, the world will be exposed to destruction that no one now can imagine or delineate. Because of secrecy, few people have sufficient information to discuss the problem intelligently, and even those few cannot predict how the techniques of the atomic age will develop or what controls will be necessary a few years hence.

The report prepared by the board of consultants to the State Department is the first ray of hope that the problem of international control actually can be solved. The report proposes a bold and dangerous solution; but inaction and an unplanned drift into international competition would be still more dangerous. Instead of searching cautiously for a minimum of action, the board proposes an aggressive policy of planned progress and international cooperation.

The recommendation of the board is the following: An international Atomic Development Authority (ADA) shall be set up with a dual purpose—to aid in developing peaceful applications of atomic power and to devise and administer a procedure aimed at prohibiting military use of atomic energy. The ADA will prospect, mine, and refine uranium and thorium; it will enrich U-235; it will produce plutonium in power plants to be equitably distributed throughout the world; and it will carry out research and development in atomic explosives. No nation or individual will be permitted to engage in any of the above activities. In other fields connected with nuclear physics, independent work by nations, research institutions, and individuals should be encouraged. For that purpose, the ADA will make available radioactive materials for scientific, medical, and technological work. It will lend uranium-235 and plutonium for use in small nuclear reactors, which will serve as sources of neutrons and gamma rays and will be most valuable research tools. Those reactors will be run at a low power level and therefore will be unsuited to the production of bomb materials in dangerous amounts. Uranium-235 and plutonium will be also loaned, under certain safeguards, for use in units producing power in significant amounts. These larger units are dangerous, because they are in principle suited to produce atomic explosives in significant quantities.

The following safeguards against misuse and evasion are contemplated:

1. It will be difficult to prospect or to mine uranium and thorium secretly if the ADA engages in this work on a worldwide scale. Without those raw materials atomic explosives cannot be produced in large amounts, not even with the help of an atomic power plant.
2. Designs for nuclear reactors and power plants operated by national governments or individuals must be approved by the ADA. The approved designs will be such as to make it impossible or at least difficult to use the plants for production of additional nuclear explosives.
3. Uranium-235 and plutonium will be loaned in a denatured form, in which they are "unusable by any methods we now know for effective atomic explosives." One can develop methods for removing the denaturant, but that requires "scientific and engineering skill of an appreciable order."
4. All reactors and power plants will be inspected from time to time. The inspection may take the form of cooperation in engineering and research between the ADA and the men who operate the reactors and power plants.
5. Through the varied scientific activities of the ADA, contacts will be established with most interested scientists and engineers in each country. That in itself will make secret activities in the relevant fields quite difficult.

Excepting the control of raw materials, no systematic or large-scale inspection is contemplated. The authors of the report are of the opinion that an extended or rigid inspection is unworkable, for it is a police measure which would engender a desire for evasion. It would not attract the high-class inspection personnel without whose ingenuity the inspection could not work. The proposed scheme, on the other hand, is a positive program in which the best-qualified talent of each country would gladly participate.

The report also contains a discussion of the transition period during which the plan is to be put into effect. At the present, the United States enjoys an advantage over other nations. That advantage consists of knowledge, experience, stockpiles, and plants. The ADA will evolve a system in which no one nation has such advantages.

It is proposed that this should happen gradually as the ADA develops. For any advantage that we give up, we shall be compensated by the progress of international organization and security. Detailed know-how, stockpiles, and plants need not be shared in the initial phases of planning. Theoretical knowledge must be made available at an earlier date because without some general knowledge the organization cannot be set up in an intelligent way. If the international organization should break down, the disclosed information may shorten the time a competitor would need to develop atomic bombs. It is claimed in the report that the time may be shortened by not more than a year or so.

The technical facts or assumptions on which the report is based are:

1. Theoretical and scientific foundations of atomic power are well enough understood so that we know in what direction to look for further developments. Thus an organization which itself is very active in research and development can keep abreast with discoveries and maintain intelligent leadership and control.
2. All exploitation of atomic power must start from the raw material, uranium. To some extent uranium can be replaced by thorium. Rich uranium and thorium minerals are not widespread. Complete control of uranium and thorium production seems, therefore, feasible.
3. Nuclear reactors and power plants can be so constructed as to make it difficult to use them for the production of atomic explosives. If explosives are produced, this can be detected rather easily by an experienced person having unrestricted access to the power unit.
4. Uranium-235 and plutonium can be so denatured that they keep their usefulness in power-producing units but cannot be used in atomic bombs. Removal of the denaturant is difficult.
5. Disclosure of purely theoretical information will speed up a potential competitor by not more than one year.

Although none of those statements is absolutely certain, all but the last two are well founded. For instance, the prospect that the ADA will be able to keep up with the development of the art is a probability and not a certainty. But no progress could be made in the difficult question of international control if only foolproof arguments were admitted.

Assumptions 4 and 5 deserve special comment.

The report's statements about denatured materials seem over-optimistic. Denaturing is certainly to be recommended as one of the several safeguards that will make violation of international control more difficult. But denaturing does not make the use of materials in atomic bombs absolutely impossible; it only impedes such use.

The assumption that disclosure of theoretical information will speed up a competitor by not more than a year is plausible, but it depends on so many intangible factors that one wonders how the board ventured to make so definite a statement. Our theoretical investigations have helped us to avoid many pitfalls. Can we be sure that our competitors will make mistakes and thereby lose time, unless they are warned by us? Are we sure that they do not now have information which, if added to the knowledge we can transmit to them, will produce practical results much faster than we may currently imagine?

That criticism, however, is directed only against minor points in the report. The main facts presented are certainly correct, and the suggestions made are ingenious, daring, and basically sound. It is the right decision to make materials available for important research and development projects, even if those materials are dangerous. By taking this risk, one will have gained more than one has lost. If such a liberal policy is pursued, people will be less desirous to break or circumvent international agreements.

If the constructive and imaginative spirit of the State Department report is compared with the "Maginot-line" mentality of "keeping the secret," one can hardly doubt in which direction our eventual hope for safety lies.

There remains one serious doubt in my mind. Is the control proposed in the Acheson report sufficiently effective? Should one not try to vest the Atomic Development Authority with more concrete powers?

I recommend consideration of the following additional proposals:

1. Every country should be permitted to send to any country as many agents as it pleases. The agents would be nominated by the country they represent and approved by the Atomic Development Authority. Their number would be determined by the country they represent, and their expenses would be charged to that country; but they would be responsible only to the Atomic Development Authority. The agents should have the right freely to inquire into any activity that may seem to them directed against their own country, or against world peace.
2. It should be considered the duty of every citizen of every country to give full information to the agents of the Atomic Development Authority. International law—superior to any national legislation—should protect men who have given such information.

One consequence of this proposal is that as soon as it becomes effective, all secrecy of information must cease. It therefore may be argued that we could be giving away more this way, and at an earlier time, than is proposed in the Acheson report. I believe that we actually shall lose little and gain most important advantages.

By giving full information to all comers, we shall not lose our most essential advantages in atomic power. We shall retain our present installations, and we shall retain our experience in production. The latter cannot be given away except by a process of education that is likely to take years. The real "secrets" are exactly those production procedures that cannot be communicated readily but must be learned by experience.

On the other hand, we shall have created an atmosphere of completely free discussion. Only in such an atmosphere is it possible to start the joint enterprise of exploiting atomic energy with full vigor and confidence. If information were given away piecemeal—as suggested in the Acheson report—we would prolong the feeling of uneasiness and mutual suspicion.

At the same time a great number of freely circulating agents would make evasions more difficult. Thus we could be more certain that the agreement would continue to function.

The effect of the proposed measure would be to place a considerable group of men directly under an international body and to protect freedom of information by supreme international law. Thereby we would have taken a first step toward placing authority in the hands of an organization whose essential function is to keep the peace. Only such a central authority can bar the road to power politics and help us to find the way to world unity.

Finally, the agents of the ADA would be in the position to start breaking down the barriers that now separate nation from nation. They should not, in their usual function, be considered policemen. They should work for world unity, and they must try to remove reasons for friction—both material and spiritual.

One will not gain real confidence in the stability of the world structure until tyranny has disappeared from the earth and freedom of speech is ensured everywhere. To reach that goal may not be feasible in the immediate future. If the present proposal is put into effect, at least this much will have been achieved: We shall have a way to protect a man who has raised his voice for the purpose of safeguarding peace.

12

Testimony on the Nuclear
Test Ban Treaty

*   *   *

Author's Note. After the failure of the Baruch plan, several years elapsed before the first attempt at negotiated arms control met with any agreement. With the rapid development of more powerful attack weapons, it is not surprising that government officials looked at possible ways to slow or stop that development. In 1958, an informal gentleman's agreement to stop all nuclear testing went into effect. In spite of the Soviet breach of the agreement, a new formal limited test ban treaty was proposed for ratification in 1963. The edited and abbreviated version of my testimony on the agreement, which follows, neither adds any information to the original nor omits any from it.

STATEMENT TO THE SENATE COMMITTEE ON FOREIGN RELATIONS,
AUGUST 20, 1963

In January 1939, I first heard about nuclear fission. Within a few days, the grave consequences of that discovery were clear to most of us who had been present at the announcement. Concern about those consequences has not left me since that time. The intervening decades have been full of surprises. At no turn did any of us know what the next step would bring.

In order to put the future into perspective, it is helpful to consider a little of the past. During the war many of us labored with diligence and success on the first nuclear explosion. The result of our efforts was a surprise. The day the first nuclear explosive was tested, no prediction had succeeded in guessing the power of the explosion. After four years of strenuous effort and theoretical calculations, we had not succeeded in predicting what would happen. The accurate knowledge gained from the experiment could not have been replaced by any amount of prediction.

Most of us believed, and the U.S. intelligence service firmly and unequivocally predicted, that the Soviets would not have a nuclear explosion for many years. The first Soviet atomic bomb test in 1949 was a complete surprise.

Some of us got very worried about the next surprise that might be in store for us. So five years after Hiroshima, we started work on the next step, a step that the majority of scientists predicted could not be done—a thermonuclear explosion. Within a short time, we produced an explosion a thousand times as great as the atomic bomb. Within an even shorter time the Soviets duplicated our result. Both of those explosions contradicted most expectations.

Right after the war it was believed that missiles could not be targeted accurately enough to justify carrying the comparatively heavy nuclear explosives of limited yield that were available at that time. When thermonuclear explosions succeeded, one could no longer avoid the knowledge that missiles would be of deadly significance. But even during the time that General Shriever was making his fantastic contribution in developing our missile systems, we still did not realize the potential that rockets would open up in space exploration, nor did our intelligence service alert our government to the Soviet activities in that area. In 1957, Sputnik took us by surprise once again. We entered the missile age with inadequate preparation. We had failed to foresee the future.

The United States held its last big well-planned test series the following year, in 1958. The missile age was still young, and the experiments then performed were not directed sufficiently toward preparation for the new developments on the horizon. The explanation lies partly in the fact that we had succeeded in our clear objective of making explosives both powerful and light. There was, however, another point we neglected.

In 1958 none of us considered missile defense a realistic possibility. In that test series, we studied neither the effects that are of primary importance in stopping a missile, nor those that are relevant to the defense of our missile sites against incoming missiles. That same year, 1958, we went into a prolonged moratorium on nuclear testing. We tried hard to foresee what was going to happen, but we did not succeed.

In one respect we took an extremely important step. In 1957 we made our first underground test of a small nuclear explosion. I argued for it mostly on the basis that this scientifically unknown area should be explored. We learned a lot from the test, including something about the possibility of clandestine testing, an area of inquiry far from my mind when I proposed the experiment. In 1958, after the United States sat down at the conference table, it became increasingly clear that while big atmospheric explosions can be policed, there is some doubt about effective policing of small atmospheric explosions and serious doubt about explosions underground. We knew this only because we happened to test an underground explosion in 1957.

Our next surprise has been better publicized. In the late summer of 1961 the Soviets resumed testing. We know now that the Soviet test series included by far the greatest megatonnage ever exploded. We have reason to believe that the series was the most comprehensive in the whole history of Soviet testing, and the most thorough carried out by any nation to that date. I don't think any expert will disagree with me when I say that this test series had to be prepared many months in advance, and that the preparations had to be extensive, widespread, and should have been open to intelligence operations. Yet on the day before Khrushchev made the announcement that the Soviet Union would resume testing, we still did not know a test series was impending.

The Soviets had another surprise in store for us. That surprise was their announcement that they had made strides toward missile defense. In 1961, and in the similarly impressive Soviet test series of 1962, the Soviets had every chance in the world to make the observations in the atmosphere that are crucial to planning an effective missile defense.

A few years ago I firmly believed that missile defense was hopeless. I am now convinced that I was wrong. Stimulated by the Soviet statements, we looked into the matter very thoroughly. I am now convinced that we can put up a missile defense that can stop the attack of any weaker power, such as China, for the next two decades. In a time when we rightly worry about proliferation, we must not neglect our defenses against an attack from a quarter other than the Soviet Union.

In addition I also believe that our defense can be partially effective against the Soviets. We may not be able to save our cities, but we may be able to save our retaliatory capacity, and thereby we may lend force to the argument that we can hit back. So long as that is true, the Soviets are unlikely to hit first. We can make a missile defense, in all probability, that will safeguard at least reasonably strongly built shelters for our people, which can and should be built.

No matter what we do about missile defense, I still believe a nuclear war will be very terrible. But missile defense may make the difference between the end of national existence and survival as a nation. By deterring the Soviets, missile defense may make the difference between peace and war.

Returning to the proposed test ban treaty, I want to review the main arguments for and against the treaty. The strongest argument for the treaty is the claim that this treaty will stop the spread of nuclear weapons to other nations. Secretaries Rusk and McNamara, in testimony, have spelled out how the treaty would prevent proliferation simply and clearly: Any nation that signs the treaty will have to test underground, and underground testing is more expensive.

On that point, there is a simple statement connected with dollars about which there is no doubt. An underground test of a magnitude traditional for the first test of a nation will cost approximately a million dollars. That is the cost of testing. It does not include the cost of the weapon. It does not include the much greater cost of developing nuclear reactors or some other source of bomb material. It does not include the cost of the research that went into constructing the bomb.

The million-dollar figure for underground testing is slightly higher than the cost that would be incurred if the test were performed in the atmosphere. But no matter how the two costs compare, once a nation has gone to the expense of developing a nuclear explosive, the additional single million dollars needed for underground testing will certainly not be a financial deterrent.

I had an opportunity, a few days ago, to make the same statement to Secretary McNamara. He answered that he did not mean the cost in dollars. He meant the psychological and political cost. I am glad to concede that under this test treaty, the Swedes and the Swiss may refrain from underground testing in order to be as completely in agreement with world opinion as they can be. I doubt whether more dangerous nations will be deterred.

The main argument against the treaty is that by permitting nuclear explosives to be tested underground, the development of the most important kinds of aggressive weapons will continue unabated. The treaty will have no effect of slowing down the development of the instruments of mass destruction. What this treaty will prohibit is the acquisition of knowledge about effects of weapons. That is particularly true of those effects that are of vital importance to ballistic missile defense.

In the early 1950s, we made plenty of observations of the effects of nuclear explosives on houses and ships. But in those days we were not considering missile silos or investigating how nuclear explosives might be used to stop a missile attack. Since the missile age began, we have conducted only one test series in the atmosphere. That was performed in 1962, just after the Soviets breached the moratorium. The series was put together in a great hurry, and the test results left us with the conviction that the amount of knowledge we desperately needed vastly exceeded the knowledge that we possessed. By contrast, in the missile age the Soviets have tested in an essentially unrestricted manner.

Secretary McNamara has told you that the Soviets probably have no more information on missile defense than we do. He has not mentioned the source of his information. Some of the Soviet experiments in 1961 were unmistakably designed for missile defense. They were few, but many other experiments may have included apparatus that looked at effects relevant to missile defense. Our information about Soviet nuclear explosions is very limited. What we know firmly is only the great extent of our ignorance. What the Soviets may have learned about missile defense in more than one hundred atmospheric explosions, we have no better way of knowing than we had of knowing whether they were preparing a test series in 1961.

On the basis of the past performance of our intelligence service, we should not assume that we know what the Soviets know. What we do know is that they have had three or four times more opportunities to find out the relevant facts about missile defense than we have had.

Secretary McNamara has told you that the United States is stronger than the Soviet Union, that we have many more missiles, that when they build more missiles we will know it and can outbuild them any time. Perhaps our intelligence service has improved enough that we can rely upon it. I hope so. But I must say that of all the areas in which it is difficult for a democracy to compete with a police state, intelligence gathering is likely to be the most difficult.

This test ban has nothing to do with how many missiles either side builds. If the Soviets want to build a huge missile force to attack us, they can do so legally under the proposed treaty. This test ban does have something to do with limiting the knowledge, but it has very little to do with limiting the knowledge needed to develop aggressive potentials further. It has something dangerously important to do with limiting knowledge concerning missile defense and concerning the vulnerability of our retaliatory forces. I believe that the Soviets have acquired such knowledge.

I believe that, because they have acquired the knowledge, they don't need any more atmospheric tests. I believe that is why Khrushchev is willing to sign the treaty. In 1960, he wasn't willing to sign, but now he has had the magnificent test series of 1961 and 1962. He now knows how to defend himself. He now knows, probably, where the weaknesses will lie in our defense. He has the knowledge, and he is now willing to stop and prevent us from obtaining the same knowledge he has. What is needed is knowledge, and that is what the Soviets have. Knowledge is what we don't have, and the acquisition of that knowledge is what this treaty will prevent us from accomplishing.

Secretary McNamara has told you that if the United States doesn't have enough knowledge about the hardening of our missile sites, we will make up for it by building more missiles and by spacing the missiles farther apart. All this costs money. All this costs billions of dollars. What Secretary McNamara is telling you is that he is willing to substitute brawn for brain, that he is willing to spend more and more money for arms. That process is rightly called an arms race. Acquiring more knowledge in order to defend ourselves cannot properly be called an arms race.

This treaty will not prevent the arms race. It will stimulate it. This treaty is not directed against the arms race. This treaty is directed against knowledge, our knowledge.

Secretary McNamara correctly stated that we can do a lot about missile defense. We can study incoming missiles and decoys, and try to see the difference between them. We can perfect our radar systems and make them harder, more versatile, and faster. Through underground testing, we can develop the best kind of nuclear explosives for use against an incoming missile. Nuclear explosives are needed when shooting at such fast and uncertain targets. A small nuclear explosion is an adequate substitute for a nonnuclear rocket with bull's-eye accuracy. But missile defense is the most complex military operation that I have ever considered. Trying to build up our defenses without proper experimentation is most hazardous.

Some of you may have read the recent book *Military Strategy* by Marshal Sokolovsky, which has been translated into English. The

Soviets were hurt badly at the beginning of World War II. Sokolovsky emphasizes the importance of air defense and missile defense. We are still hesitating about a missile defense. The Soviets are working on it. With all the uncertainties that will continue if this test ban is adopted, we may never deploy a missile defense. That is why I say that this treaty is a step in the wrong direction.

In regard to missile defense, the question of our relationship to our allies must be raised. I believe that cooperative research on defense by nuclear means can proceed without inviting war by accident or risking misuse of these weapons. We can tie missiles to their sites and defensive nuclear warheads to the missiles. We can put an electronic program into the whole system, an electronic program that will prevent a missile from exploding before it has reached an altitude where it can do no damage on the ground. We can make sure that the missile will explode or otherwise be destroyed before it leaves the territory of the country that fired the missile. Such missiles cannot be used for anything except to shoot down flying objects. They can be used for only one purpose—for justified self-defense.

By working together with our allies, through common work and common results, we could strengthen our alliance. Indeed, in order to save our alliance, we must make the defenses common. Yet it seems that the proposed treaty, unless very thoroughly amended, will erect a big barrier between our allies and ourselves in the most important area of nuclear defense.

I would certainly prefer that if a treaty is to be ratified, it should restrict nuclear test explosions and nothing else. We must maintain our right to use nuclear explosions in our defense and in the defense of our allies, in defense against massive aggression.

There is another surprise around the corner. The surprise will not be so great, because some of us have foreseen it and worked for it for years. Nuclear explosives can be used for peaceful purposes, for moving great amounts of earth at one-tenth the conventional cost. We can build harbors, dig sea level canals, deflect rivers, throw off overburden from deep mineral deposits, and thereby increase our wealth and the wealth of other nations. The explosives can operate in a very clean way so that no one will be exposed to more radiation than is experienced year in and year out without effect in our laboratories.

There is some measurable radioactivity connected with such projects. The proposed treaty prohibits the deposition of any radioactivity outside the territory of one's country, which means that the United States could not use such explosives to help a poor country to develop. The Chairman of the Atomic Energy Commission, Glenn Seaborg, has rightly emphasized the great importance of the peaceful use of nuclear explosives called Plowshare. He has rightly stated that the proposed treaty does not prohibit our developing Plowshare inside the United States as long as no radioactivity is deposited elsewhere.

Seaborg also has said that when that capability is fully developed, we will have to go back to the Soviets and ask for a change in the treaty. I am a little doubtful as to whether the Soviets will be willing in the future to give up the hold that this treaty gives them over our negotiations with friends and other countries. The treaty explicitly prohibits nuclear explosions in space; yet many of the possible and as yet unproved applications of nuclear explosions are in space exploration. Space propulsion by means of nuclear explosions may become possible.

The main consequence of the proposed treaty will be to limit the acquisition of knowledge that we need for defense. It also will interfere with the application of such knowledge as we may acquire in the future that would assist in peaceful purposes in science or in a space adventure.

*   *   *

Author's Postscript. When I offered this testimony in 1963, I did not touch upon one argument that had great popular appeal: Radioactive fallout from atomic tests would be stopped. At that time, the radiation exposure from the atmospheric tests was less than 2 percent of the exposure to radiation from natural sources. Those sources include cosmic rays, naturally occurring radon and uranium, and a multitude of other sources that have been present far longer than mankind. At the time of the limited test ban hearings, methods were available to limit the radioactivity from a nuclear explosion. Because the radioactivity from atmospheric tests decays comparatively rapidly, a continuing decline in test radioactivity could have been easily demanded and enforced.

On the other hand, I failed to mention an important argument against the treaty. Atmospheric experiments produced a great deal of information about the nature and direction of Soviet progress. With the onset of underground testing, that source of knowledge dried up completely. The Soviets, however, can rely quite successfully on straightforward espionage.

One of the arguments I stressed—that the United States research on defense hardly could be accomplished without opportunities to test in the atmosphere—proved to be overstated. Ingenious underground experimentation was more

13

The Beginnings of Secrecy
in Physics

When the George Washington University conference on theoretical physics opened on January 26, 1939, Niels Bohr made the first announcement: Otto Hahn and Fritz Strassman of the Kaiser-Wilhelm-Institut in Berlin had discovered fission. Within hours, the news had spread through the world of physicists like wildfire.

The discovery of fission was indeed a discovery in the literal sense of the word. All of the facts had been there for a few years. What was required was for someone to notice them, but understanding was delayed in a remarkable manner. At the same time, when it came, the discovery was both important and exceedingly dangerous.

Those of us acquainted both with Hitler's policies and Germany's excellent physicists had an immediate concern: the possibility of a weapon based on the release of nuclear energy. The steps by which fission might be harnessed might be invisible to the layman's eye, but it was impossible that the prospect would not be as clear to the excellent German scientists as it was to all of us.

Scientists who would never have considered, much less tolerated, secrecy in science suddenly began to consider the means by which work on atomic physics could continue without publication. The outstanding exception was Niels Bohr. As in many other respects, he had his own sharp and unwavering opinion. He continued to advocate a policy of openness. That attitude, however, seemed to ignore the obvious danger of Hitler's acquiring an atomic bomb and using it to conquer the world.

So in early March, Leo Szilard, Johnny Wheeler, Eugene Wigner, Victor Weiskopf, and I met with Niels Bohr in his office at Princeton, where he was visiting. Our purpose was to convince Bohr that while fission research should continue, no results should be published lest the Nazis learn from them and produce nuclear explosions first.

Bohr did not like the idea. One by one, Bohr took us aside and, running along the corridors of the university, whispered (his customary means of communication on questions he considered important) his concerns. "Joliot [a famous French physicist and son-in-law of Madame Curie] would not like such secrecy. He is already unhappy not to have gotten credit for the discovery of the neutron, and he has been working on fission."

What an argument! World war and conquest by the Nazis were on everyone's mind, and Bohr seemed concerned, in a thoroughly unselfish way, about who would publish first. What really moved him was the old code of behavior in the scientific community. For him, a great relativist, that was an absolute standard.

Bohr finally agreed to send a wire to Joliot suggesting he not publish results on fission. However, according to an unverified story, because the telegram arrived on April 1, Joliot considered it an April Fools' Day joke and therefore published with no compunctions. The two verifiable facts are that Bohr did write asking Joliot to hold up publication, and that Joliot did publish his findings on fission.

After the meeting at Princeton, I had been delegated to talk with Enrico Fermi, who also opposed secrecy. Fermi had just started work at Columbia, having used his Nobel Prize money to escape from fascist Italy. Reluctantly, Fermi agreed not to publish his findings, provided everyone else did the same. When Joliot published, Fermi also resumed publication. But the change in attitude toward secrecy in science had begun. Soon thereafter, the government introduced comprehensive secrecy practices.

Today, secrecy has become a terrible destructive force in our society. My postwar efforts to reverse the process have not affected its devastating spread. I am unhappy that I had anything to do with its beginnings.

The last time I met Niels Bohr was in Geneva in 1958 at the Second Atoms for Peace Conference. He arranged for the two of us to have dinner together. Throughout the evening, Bohr (exclusively) talked about the need to get rid of secrecy. His arguments were wasted, because I had long since agreed with him. The next morning I was the first speaker. My talk was the result of years of effort on behalf of opening up fusion research to international scientific discussion. That was my only success in actually making inroads against secrecy.

Throughout my talk, Bohr sat in the front row with his eyes closed. It was not compatible with the uncertainty principle to discover whether or not he was asleep.

14

The Road to Nowhere

Science thrives on openness. Researchers should, and often must, share their findings. But during World War II we were obliged to put secrecy practices into effect.

After the war, the question of secrecy was reconsidered. Niels Bohr, the greatest physicist of the century, suggested: "In the cold war each side should use the weapon that it can handle best. Secrecy is the appropriate weapon for a dictatorship; openness is the weapon democracies should use." Bohr was dismissed as a crazy scientist when he carried that message to President Truman and to Winston Churchill. They decided that only information that does not endanger national security should be published. The result was a proliferation of secrecy in science.

After many months of political inaction following Hiroshima, the United States presented the Baruch Plan to the United Nations: The United States would share the secrets of nuclear energy in exchange for all nations agreeing to international control of nuclear power. The plan failed—the Soviets were not interested. That is not surprising, considering that they had probably already guessed our atomic secrets. As I wrote in an atomic alphabet for my young son:

S stands for secret; you can keep it forever.
Provided there's no one abroad who is clever.

But the practice of classification continued; it was our "security," whether it worked or failed.

In 1954 President Eisenhower proposed an international conference to explore the potential peaceful uses of atomic energy. Whether the Soviets participated or not, we would share the information. To do otherwise would have made little sense since scientific secrets do not keep.

We gave away a lot of information at the First Atoms for Peace Conference, and we accomplished a lot. Soviet scientists were delighted to present their achievements; clearly, an earlier refusal to participate in the conference was made by politicians, not scientists. With secrecy on reactor designs lifted, schools of nuclear engineering were established. Industries hired the graduate engineers, and a dozen years later nuclear reactors competed with coal, oil, and gas in generating electricity.

The balance of benefit and loss in the first two decades of the nuclear "security" system seems clear: Decisions made with inadequate preparation, self-deception, and diplomatic failure prompted the deterioration of cooperation in the free world. Security regulations have also helped drive a wedge between our universities and our military research and development effort.

Under present rules, research done in our national laboratories cannot be fully shared with civilian industries. When we fail to expose people to problems they could help solve, we remain unaware of the loss. We now have millions of classified technical documents. We also have falling productivity. Rapid progress cannot be reconciled with central control and secrecy. The limitations we impose on ourselves by restricting information are far greater than any advantage others could gain by copying our ideas.

In addition, by tainting science with secrecy, an unfortunate public attitude is perpetuated: Science is nobody's business but the scientists'. Today, science and technology are part of the life-support system of the world. Encouraging the development of a scientifically literate public is of primary importance to everyone's well-being.

Secrecy is not compatible with science, but it is even less compatible with democratic procedure. Two hundred years ago James Madison said, "A popular government without popular information, or the means of acquiring it, is but a prologue to a farce or a tragedy, or perhaps both." The term *credibility gap* is a modest description of our monstrous current problem. Of its many secrets, the government has controlled none more rigorously than information about the growth of Soviet weaponry and technical military expertise. But without such knowledge, the question of how much and what kind of defense our nation needs is left to the shifts of political winds.

How can we dismantle this expensive, gargantuan system of classification? First, we should identify those matters that we quite properly should try to keep secret. We do not want press coverage of delicate diplomatic discussions or premature disclosure of government plans.

But information pertaining to science should be more readily available. For the last century, American industries have had *company-private information*, which allows dissemination of general knowledge but not the delivery of blueprints to one's competitors. It works because it restricts only the tricks of the trade—ideas that are hard to transfer except by detailed explanation. Information on basic issues remains free.

A similar system should be applied to research in national laboratories. I propose that a time limit be imposed on the classification of information as secret, top secret, special access, "burn before reading," or whatever. Technical advances pertinent to defense should be classified for no more than one year and should then be shared with industry on the basis of company-private procedures. Expert supervision could ensure that the few cases in which longer-term secrecy is necessary are strictly limited.

Although we have tried for years to achieve greater stability in international relations through arms-limitation agreements, the results have not been encouraging. The most important work today is on weapons of the future. As long as such arms are secret, they cannot even be discussed, much less limited. Adopting a policy of openness as the first and only unilateral step toward disarmament would strengthen our relationships with our allies as well as illustrate the advantages of freedom as compared to the practices of our Soviet colleagues.

One of the earliest tales I remember from my childhood is about the adventures of a band of runaway puppets. They reached a fork in the road where one signpost pointed to "Nowhere," the other to "Everywhere." The puppets chose the latter, got into trouble, but in the end got safely home. I do not claim that openness will never lead to trouble, but I am sure that it offers us the best possibility of getting safely home.

15

Ernest Orlando Lawrence

LLNL stands for Lawrence Livermore National Laboratory. Ernest Lawrence created it.

In the midst of the hydrogen bomb controversy, I left Los Alamos. I had become convinced that at least two laboratories were needed to maintain excellence in government-funded, secret research. The technical criticism of the scientific community was lacking. Competition from a second laboratory was needed to provide comparison and stimulus.

In 1952 Ernest Lawrence, who had been instrumental in getting work on the hydrogen bomb under way, called me and invited me to Berkeley. On my arrival, he took me to a famous San Francisco restaurant, Trader Vic's, the place where he conducted all his really serious business. He told me he was ready to start a second government laboratory to do research on atomic weapons. He proposed that I argue for the idea in Washington. I did not think I could do it, but that is just what I did in the next few months. To my surprise, both the Department of Defense and the Atomic Energy Commission agreed to establish a new laboratory on nuclear weapons in Livermore, California.

At that point two of my very close friends—Enrico Fermi and Johnny von Neumann—told me that I must not go to California and join the new laboratory at Livermore. Those two were among the greatest scientists of our century. They were also two men whose friendship I was sure to retain regardless of what I decided. Never before had they offered me unsolicited advice. On that occasion, they did. But I did not take it.

Ernest Lawrence told me I was needed in Livermore, and my wife wanted to go to California. But independent of those circumstances, I was convinced that I had to make a contribution where others, who could have done so, were unwilling to make the necessary effort. I turned away from my original choice—to work on pure science—with my eyes open. What I could not foresee was that within a couple of years both Enrico Fermi and Johnny von Neumann would be dead, and within five more years Ernest Lawrence would die. None of them reached the age of sixty.

During the early years of the new laboratory, Ernest worked tirelessly. He recruited most of the early workers. Several of his former students, Herbert York, Harold Brown, and John Foster, became directors there. His support never wavered, even when there were reasons to question the laboratory's performance. The first test, which we conducted in Nevada, didn't work. The second test, conducted in the Pacific, didn't work either. The third test, again in the Pacific, was scheduled but had to be postponed because of weather conditions.

The information from the test failures came back to the Livermore Laboratory. Needless to say, we went over it most carefully. One of my friends, Montgomery Johnson, finally suggested the reason why we had failed and why the next test was going to fail.

Ernest visited Livermore. On our ride back to Berkeley he asked me whether the people in Livermore were sufficiently qualified. I assured him that they were. In a difficult field we had made a mistake; as a consequence, the third test that was to be carried out would fail. Ernest recommended that I fly out myself and stop the test.

Stopping a test is not easy. All the preliminary equipment had been set up, and everything was ready except the weather. Still, we were spending a lot each day while waiting for the test.

The director of the lab, Herb York, was in charge. I arrived on the atoll and met with him. I went through the explanation of the difficulty, and finally York agreed. Then Harold Brown, the second in command, came along, and I went through my routine again. Ernest was correct in his approach. The decision to cancel the test was made. The men in charge took full responsibility. From then on, Lawrence's laboratory had successes.

The beginning history of the Lawrence Livermore National Laboratory illustrates a point that Niels Bohr liked to make about expertise. Bohr defined an expert as "a person who has, by his own painful experience, found out about every conceivable error in a very narrow field."

But the problems of the laboratory did not disappear when we gained our expertise. After about four years Herbert York, the first director of the lab, left for a new job in Washington. Two weeks later he changed his mind about what the laboratory should be attempting to do. York told me about his change of mind. That was a help. I took over the directorship of the laboratory in order to work toward ensuring that it would continue innovative work on nuclear weapons.

Within a few more months, Ernest called to ask me to go to the talks being held in Geneva. A moratorium on testing was under consideration. I refused for a number of valid reasons. Ernest decided to go himself.

For many years he had suffered from ulcerative colitis. I had developed the same disease shortly after we moved to California, and by coincidence we saw the same doctor. When Ernest learned of my illness, he immediately came to my office. His instructions were excellent. "Do what the doctor says. Don't get excited. Don't work too hard."

Unfortunately, Ernest did not take his own advice. When he left for Geneva, he had gotten over his most recent attack of colitis. He was optimistic about what the talks in Geneva might accomplish. Three weeks later he was flown home to the hospital. I saw him for half an hour. We talked about the future of the laboratory. A few days later he was taken into surgery. He never recovered.

Ernest had a profound influence on the lives of many people—on graduate students, on his collaborators, and on General Groves, the leader of the Manhattan Project. Although I worked with him for less than ten years, he had a deep influence on my life. His persuasiveness stemmed largely from his own dedication. The laboratories he fostered stand as monuments to his work. I was honored when his brother, John O. Lawrence, a fine physician, asked me to give the first E. O. Lawrence Memorial address in 1960. The essay that follows is adapted from that speech.

16

The Hazards of Radiation

Ernest O. Lawrence changed the direction of modern physics. He was a builder, and the influence of his work reached from the investigation of the nucleus to national defense to medicine. He developed the first cyclotron, an instrument for accelerating particles to probe the atomic nucleus. In that development, he stands above everybody. He promptly applied the cyclotron to medical purposes. For many years the Berkeley Radiation Laboratory was the only source of radioactive isotopes, a new means for conducting biological research. Because of his initiative and enthusiasm, our country enjoyed a considerable head start in nuclear medicine.

Two applications of isotopes in medicine are obvious. One concerns our fight against one of the greatest remaining dangers to human health: cancer. The other use is even more important. With the help of isotopes, we can follow the way in which an element goes through the complicated maze of biological activities. Such observations provide detailed and instructive insights into biochemical processes. But the first concern was to determine the hazards of radiation.

Ernest's experience in evaluating radiation hazards began in 1935, when the 37-inch cyclotron first started to operate, producing neutrons. One obvious question was whether the biological effects of the neutrons would be similar to the effects of x-rays, gamma rays, electrons, alpha particles, and other types of radiation. Because that question arose, Dr. John Lawrence and Dr. Paul Abersold rigged up a little apparatus with a rat inside. The device had to be jammed into a narrow space in the cyclotron. The rat was supplied with air through a little tube. The cyclotron was turned on. It ran for a few minutes. Ernest said, "We'd better stop and see how the rat is." He looked, and the poor rat was dead.

That caused consternation, because the rat had not received a very big dose of radiation. It seemed that neutrons were much more dangerous than other forms of radiation. It took a little time to discover the real story: The tube that supplied the air to the rat had been pressed shut during the test, and the rat had suffocated. That was the first of many alarms about radiation. In this case, alarm led to a positive result. It made all the people in the Radiation Laboratory conscious of the possible danger of radiation. Partly because of caution and partly because of good luck, the Radiation Laboratory has a really excellent safety record.

Starting from those early observations, an important conclusion developed. Nuclear radiation has a straightforward effect: It breaks molecules apart. There is nothing particularly specific in this. Some types of radiation are two or three times more effective in fracturing molecules than some other types. But by and large all forms of radiation act in a similar way when they make contact with tissues. The main question is how much energy is delivered and to which tissues.

Of course, the overall effect may appear quite different. Penetrating radiation traverses the whole body, while soft radiation stops in the skin. Radiation from isotopes may be deposited in certain organs, and only those organs will be irradiated. But as long as some tissue is irradiated, the effect of the radiation can be estimated with reasonable accuracy on the basis of the amount of radiation energy delivered to that tissue.

That is in sharp contrast with the effects of chemical agents, which have a lock-and-key property. A molecule can change from a food into a poison by means of a small change in molecular structure. To predict biological effects on the basis of chemical evidence is beyond our present knowledge and may remain so for some time. Radiation is a much simpler agent. But that does not mean its effects upon us are simple. While radiation is itself simple, we are complicated. When something happens to us, God alone knows how each individual will react.

Radiation has the same property as everything else in the world: Too much of it is dangerous. The neighborhood of a nuclear explosion contains too much of it. Therefore the immediate neighborhood of nuclear explosions must be watched carefully. It is better to do nuclear testing far from living things by putting them underground or outside the atmosphere whenever possible.

A considerable amount of controversy exists about the effects of the dispersed fallout from nuclear tests. Let me summarize the argument that fallout from atmospheric testing is dangerous. The summary is effective and simple. Exceedingly small doses of radiation result from testing by the time the radioactivity reaches distant places. Such very small doses have a very small chance of harming an individual—one chance in 100,000 or one chance in 1,000,000. But there are 3 billion people in the world, and if there is one chance in 1,000,000 then 3,000 people will be hurt, and if there is one chance in 100,000, then 30,000 people will be hurt. We should not hurt thousands of people.

A different point of view should be considered. The effect from fallout is so small that nobody has ever observed it. No one knows—not from direct observation, not from statistics, not from any valid theory—whether the claimed damages exist or do not exist. The plain fact is that we do not know the effects of small doses of radiation. A further fact is that we are unlikely ever to know because the effects are so small they cannot be identified.

The estimates that claim the lives of thousands of people might be shortened because of fallout are based on a very simple hypothesis. The hypothesis is that if 100 r do damage with a certain probability, one r will cause the same damage with one-hundredth that probability.[1]

The hypothesis is based on the single-hit theory. If one molecule is hit and disrupted, the disruption of this one molecule is irremediable and will produce an ultimate consequence with a fixed probability. We know that something like 500 r delivered in one dose will kill half the people if they are healthy, and more if they are not. Yet, we know that we can, in fractionated amounts, deliver 5,000 r of therapeutic x-rays without danger of serious short-term injury, although clear-cut danger of long-range development of disease exists. Proportionality between dosage and effect is certainly not demonstrated. In fact, there is no good statistical evidence of any damage to the individual until that individual receives approximately 50 r.

Even above 50 r, the results of experiments are complicated and conflicting; some of the evidence clearly contradicts proportionality. The evidence is derived from laboratory experiments, from irradiated populations in Hiroshima and Nagasaki, and from therapeutic applications of radiation. The upshot of those results is that no clear-cut evidence is obtained for proportionality in any pathogenic effect.

Many low-level irradiation experiments have been performed on animals. The radiation used in those experiments is one hundred times greater than the radiation from fallout. Dr. Egon Lorenz in the National Institutes of Health and Dr. Leon Jacobson at the University of Washington exposed colonies of rodents to low-level radiation.[2] They found that the irradiated animals lived 10 percent longer than the control animals, which were not irradiated. I don't know what conclusions can be drawn from those results, but the evidence for harmful effects of low-level radiation is absent.

The simple fact is that at very low levels of irradiation no one knows whether the effect is proportional to the radiation, whether it is proportional to the square of the dose, whether it has a threshold so that below that threshold there is no damage, or whether below a certain threshold there are beneficial effects. Radioactive waters used to be advertised as beneficial. That claim was unscientific. But the opposite claim that all radioactivity is harmful is not scientific either. We simply do not know.

Some effects probably are proportional to the dose received. But our bodies are complicated and do not obey simple laws of proportionality. If a thousand molecules in the whole human body are disrupted by a low-level dose of radiation and, as a consequence, a thousand cells in various parts of the body die, it seems a bit narrow-minded to ignore the body's curative mechanisms. On the other hand, if massive irradiation takes place, the body's defenses may be overwhelmed. The effects in the two situations are not necessarily any more related than the effects of a scratch and a severe stab wound.

We know that some functions of our body inhibit cancer. We know that a variety of causative agents are implicated in various types of cancer. We know that a few types of cancer—especially leukemia and cancer of the thyroid—are sometimes initiated by radiation. Apparently, the smaller the dose of radiation, the more delayed the onset of cancer. However, in no case has the proportionality between radiation dose and cancer been demonstrated, and there is no evidence of carcinogenic effects from very small doses of radiation.

The possibility of genetic changes has been discussed as widely as the possibility of cancer. In the genetic field, H. J. Muller demonstrated a clear-cut situation. His detailed studies have shown that mutation in spermatozoa—mature male genetic cells—of fruit flies is strictly proportional to the radiation dose. Proportionality has been demonstrated in a wide range from 4,000 r down to a few r. But the results are different for the spermatogonia (parent cells of spermatozoa) and for the oocytes (female genetic cells). Except for their means of propulsion, spermatozoa are bare cell nuclei; they are stripped down to the genetic apparatus.

According to William L. and Liane Brauch Russell, the effects of radiation on sperm are indeed proportional to the amount of radiation, and the single-hit theory seems to be supported. However, they found a more complicated situation for spermatogonia.[3] Spermatogonia have a cell nucleus *and* a cell body. In that case the effects of radiation depend on the rate at which irradiation occurs. The effects become smaller if the dose is fractionated instead of given all at once. Repair mechanisms seem to operate and only strong irradiation in one dose is damaging in proportion to the dose. The same seems to hold for the female cells, for the ovum. Again, repair mechanisms must be considered. In worldwide fallout the dose rate is small and spread out in time. Spermatogonia and oocytes may not be damaged at all.

The Russells have conducted so many chromosomal experiments with hundreds of thousands of mice that their work is known as the "mega-mouse experiments." The effects of mutations of low-dose radiation delivered at a low-dose rate are so small that they cannot be directly ascertained. Studies of billions of people would be required to find statistical evidence of effects of radiation of the small magnitude being discussed. The radiation effects being sought are so small that, even if they exist, they cannot be determined.

As mentioned, small doses of neutrons, x-rays, gamma rays, beta rays, alpha rays, or cosmic radiation all act very similarly. That is demonstrated by physical evidence, by chemical evidence, and by a concurrence of a great deal of biological information.

Fallout brings Hiroshima to mind. Hiroshima evokes emotions. If we want a quantitative evaluation of the effects of fallout, a discussion of cosmic rays—a practically omnipresent form of radiation—is helpful.

For more than a hundred million years, ever since fish emerged from the ocean, all of our ancestors have been exposed to cosmic rays. Actually, even when our ancestors were sheltered by many feet of sea water, they were not completely protected from radioactivity. The water contained radioactive potassium and other radioactive materials. The natural background of radiation to which our ancestors were exposed is similar in amount to what we are receiving. It is approximately fifty times as great as the radiation we get from the fallout.

In addition, we are exposing ourselves, for good reason, to medical x-rays. Consider, first, diagnostic x-rays, which for an average healthy individual amount to about as much as the natural background radiation. Therefore the amount of radiation that the average individual is getting is one hundred times as great as the fallout they are getting. Furthermore, the amount of natural radiation varies a great deal from place to place. Many people in Colorado are 5,000 feet closer to the sky. Cosmic radiation is beating on Colorado with greater intensity than on Florida, and even the difference is much greater than the radiation received from test fallout. At 40,000 feet, the average altitude of long-range jets, the effectiveness of cosmic radiation is particularly great. Those who regularly spend hours at such altitudes receive much more radiation than those who live in the high mountains.

People usually do not hear about the background radiation. Therefore, they usually do not worry about it. Yet background radiation is greater if you live in a brick house. It will be greater or less depending on one's diet. It is higher for people who live in a part of the country where there is more thorium, uranium, or radon in the soil.[4] The differences between those amounts of radiation are much greater than what we receive from test fallout.

In the 1950s the concept of the maximum permissible radiation dose was introduced. It seemed useful. It is also disturbing and misleading. What happens when the figure of the maximum permissible dose is circulated and people find that they have received 20 percent of the maximum permissible exposure? What do they think? "I have gone 20 percent of the way toward some kind of a disaster which I do not understand, but which quite possibly means getting cancer."

We should talk about one thing—the average background exposure. The average background exposure is about one-tenth of an *r* per year. The maximum permissible dose is a figment of man's imagination. The average background exposure in many cases is similar to it. Furthermore, since everybody is exposed to background radiation, it is a much more reasonable and democratic unit. It is also more reassuring. Let us discuss exposure on the basis of fractions or multiples of the average background exposure.

Global fallout from testing is not dangerous, but the fallout scare is. Many people know that medical x-rays often have a dose fifty times as great as fallout. How many people have been scared away from diagnostic x-rays? Diminished circulation to the brain can be diagnosed by injecting radioactive material into the bloodstream and watching for its appearance in the skull. Such a circulatory deficit, which can impede mental processes, usually can be corrected easily if it is diagnosed. Yet many people have gone with their ailments unrecognized and untreated, because there has been a needless and exaggerated scare about radiation.

The problem is more serious in connection with therapeutic doses. Those doses are much more massive. My sister had trouble with her thyroid gland. The tissue had to be removed either by an operation or by radioactive iodine treatment. Every operation has some small hazards. Radioactive iodine treatment is simple, painless, and much safer. My mother, however, was a great worrier. It took all my eloquence to convince my mother that the thyroid treatment was preferable. I wonder how many cases there are in which people have abstained from needed radiation treatment because of the fall-out scare.

In a serious case of cancer a surgeon may undertake an operation that is dangerous. He will tell the patient that there is a chance he will die during or immediately after the operation. In such a serious situation, it is not at all unusual to accept a 20 percent risk of death. In the case of radiation therapy, no level of hazard seems to be acceptable. I do not see any logical or sensible reason for the distinction. Either the surgical procedure is too radical or else the attitude toward the radiation procedure is too conservative. Such lack of realism can be removed only by thorough public education. The problem of explaining radiation hazards is essentially the same whether one explains the practically nonexistent hazard of worldwide fallout, the hardly greater hazard of diagnostic applications, or the real hazard of therapeutic use.

It is to the great credit of Ernest O. Lawrence and his brother John that, as soon as man-made radioactivity became available, they sought and found applications of it to biology and the art of healing. As the availability of radiation sources grew, so did the safety precautions. We should recognize the comprehensiveness of these safety precautions and exercise them, so that the harmful effects of radiation remain small in comparison to the beneficial effects. We should also not fear to pursue further development of radioactive techniques to improve the well-being of people throughout the world.

17

Widespread Aftereffects
of a Large-Scale Nuclear War

No potential war has ever been discussed as carefully and in such quantitative detail as nuclear war. None has produced more disagreement. Scientific knowledge of the aftereffects of a nuclear war—fallout, ozone layer depletion, and climatic effects—could be of great importance in making political decisions. Early fears concerning two of those, fallout and ozone depletion, turned out to be exaggerated. A third, the fear of climatic changes due to smoke and dust, is under active discussion now. A study by Turco *et al.* is an early contribution to the question.[1] The preliminary and uncertain nature of their findings deserves clarification.

FALLOUT

The consequences of war-engendered fallout on noncombatant nations were the first effects to receive widespread popular attention. During the 1956 Presidential campaign, Adlai Stevenson raised the problem of worldwide fallout in an arresting manner. In 1957 a book, *On The Beach,* and its subsequent movie version popularized and perpetuated the belief that the effects of fallout would endanger, and probably extinguish, all human life.[2]

In 1975 the U.S. National Academy of Sciences issued a report in which the estimates of war-produced worldwide fallout were so low as to have insignificant biological consequences.[3] However, estimates used in the study were scaled from larger nuclear weapons tests and therefore are not representative. Actual weapons tests avoided atmospheric conditions where rainfall was expected and were not carried out in middle latitudes. Because of their size, a larger fraction of radioactive products from the tests was injected into the stratosphere than would be from current weapons and, because of the longer residence time in the stratosphere, the radioactivity of fall-out from a nuclear war was underestimated.

The best current calculations, made at Lawrence Livermore National Laboratory (LLNL), suggest that global-scale fallout will expose people in the nonbelligerent countries in the northern 30-to-70-degree latitudes to an average external dose of 20 r units, a total dose delivered over fifty years.[4] It is important to note that this estimate holds on the average. Vagaries of weather, particularly of rainfall, can concentrate fallout in large but limited areas called hot spots, where the total dose values in nonbelligerent countries could be as high as 250 r units.[5] (The assignment of doses does not take into account the shielding afforded by houses, which would reduce the radiation received by up to one-half.)

A given radiation dose received over a long period is less harmful than the same radiation dose received in a short time, because of continual biological repair.[6] The formula in current use for slow dose rates suggests that 90 percent of the damage is repaired at the rate of 2.5 percent per day. An acute dose of 100 r units or less does not cause appreciable radiation sickness. When the biological recovery formula is applied, a dose as high as 250 r units from worldwide fallout is built up at a rate so slow that it would not produce significant casualties.

Inclusion of radioactive elements in the food chain might, on the average, increase those estimates by a factor of two.[7] Simple precautions based on generally available knowledge about fallout, such as that offered in public schools in the Soviet Union, could sharply limit the danger. The contrast between the slight long-term damage of fallout to the worldwide population and the great immediate dangers in combatant countries is obvious.

The worst case of fallout would occur if all nuclear reactors and all used reactor fuel elements were subjected to explosions violent enough to disperse their radioactive contents. In those conditions, the 20 r unit average dose and the 250 r unit "hot spot" dose may need to be multiplied by a factor of about eight.[8] Fallout in the non-belligerent Northern Hemisphere countries could then be a considerable danger. Vaporization, along with dispersion, of reactor radioactivity is not easily accomplished, nor does it appear to serve any important military purpose. It is therefore unlikely to occur.

Early underground disposal of wastes would greatly diminish the danger.

Thus, there is almost no likelihood that fallout outside the belligerent nations (with the exception of regions close to the belligerents) would have a severe damaging effect. Even in the belligerent countries, fallout would have less important effects than the immediate effects of fire and blast if the most elementary civil defense measures are exercised. Thus, fallout is far less dangerous than was originally assumed.

OZONE LAYER DAMAGE

In the 1970s, there was concern about the possible effects of a nuclear war on the ozone layer.[9] The small amounts of ozone (oxygen molecules containing three rather than the usual two oxygen atoms) present in the stratosphere play an important role in shielding the earth from ultraviolet radiation. Large nuclear explosions propel the contents of their fireballs, including the molecules NO and NO_2 (collectively referred to as NO_x), into the high atmosphere where they react with ozone, eventually converting ozone back to oxygen molecules (O_2).

If injected in sufficient quantity at high altitude, the NO_x could deplete the ozone layer significantly. Ozone molecules are steadily regenerated by solar radiation, but the NO_x residue from large nuclear explosions could reduce the effective thickness of the ozone layer by an average of 30 to 40 percent for a period of a year or two.[10] (For purposes of comparison, the effective thickness of the ozone layer during summer is about 20 percent less over Miami than over Seattle.) Complete recovery would take several more years.[11]

Assuming the explosion of thousands of megaton weapons, the calculated diminution of the ozone layer would give rise to increased serious sunburn immediately and to an increased incidence of skin cancer over an extended time. Those primary effects, like the possibility of eye damage, could be countered with simple precautions. A secondary concern is that ozone depletion and the consequent increase in ultraviolet radiation would destroy ultraviolet-sensitive plants or plankton species and would harm early springtime plant growth, with important effects on the ecosystem.[12]

Since, in middle latitudes, observed natural variations of the ozone column are comparable with the predicted effects of a nuclear war with megaton weapons, the changes are not likely to be highly significant.[13] The most pertinent consideration, however, is that the present U.S. arsenal has few nuclear explosives with the great yield that is a necessary precondition for depletion of the ozone layer. (Fireballs of smaller explosions do not rise to the ozone layer.) The same trend toward smaller, more accurately aimed, and less costly weapons also has been adopted with some delay by the Soviet Union.

The problems related to a weakening of the ozone layer seem manageable, and with the rapid elimination of weapons large enough to cause ozone layer damage, the danger of serious ozone layer depletion will disappear.

EFFECTS OF DUST

In the mid-1960s the climatic effect of dust raised by multiple nuclear blasts became a concern at Lawrence Livermore National Laboratory. (Unfortunately, the early work on the subject by R. G. Gutmacher and G. H. Higgins is still classified.) Dust in the postattack atmosphere would tend to reduce solar radiation received at the surface of the earth. If the reduction were sufficiently large and prolonged, the temperature would be lowered.

The effects produced by blast-lofted dust from nuclear explosions can be expected to be analogous to those produced by dust spewed into the atmosphere by volcanic eruptions. Annual growth rings in bristlecone pine trees (a long-lived species) show the effects of early frost, and those changes are correlated with large volcanic eruptions that distributed dust in the upper atmosphere.[14] Comparing the historical volcanic events with annual growth rings offers a rough measure of the relationship between dust and worldwide temperature.[15]

The quantity of small particles in the dust raised by a large-scale war is comparable with that produced in the largest volcanic eruptions. Recent estimates of the mass of small dust particles that would be lofted into the stratosphere in the event of a nuclear war are based on scanty information. Review and analysis of old air filter samples and recent tests involving large amounts of high explosives may provide a more reliable basis for calculations.

The most probable conclusion is that the atmospheric effects from war-generated dust would be noticeable but by no means severe on a global scale, or even in the Northern Hemisphere.

EFFECTS OF SMOKE

In 1982 Crutzen and Birks proposed that smoke particles from forest fires caused by a nuclear war would give rise to a much more serious effect than dust. Smoke could be present in greater quantities than dust and would absorb sunlight as well as scattering it.[16] In 1983 Turco *et al.* argued that smoke generated by fires in bombed cities or oil refineries would produce even greater effects than those from the forest fires.[17] In discussing those results, some scientists (e.g., Carl Sagan) have created the impression, aided by a short movie depicting the postulated effects, that smoke from fires ignited by nuclear explosions and, to a lesser extent, the dust raised by a nuclear exchange would bring about what is now known as "nuclear winter."[18]

The theory of a severe nuclear winter depends on the assumption that a large quantity of smoke will be emitted by burning cities and forests and will be distributed and stabilized in the high atmosphere. Such smoke-laden air, heated by sunlight, could raise the temperature at the tropopause (the boundary between lower, thermally agitated air masses and higher, stratified layers at about 10 kilometers) from approximately $-50°$ or $-60°$ Centigrade to about $5°$ Centigrade. At the same time, the surface temperatures of the continents in the Northern Hemisphere between the latitudes of $30°$ and $70°$ might drop to about $-30°$ Centigrade because of the absorption and scattering effect of the smoke.

A large quantity of smoke at high altitudes would produce an extremely massive inversion, a state in which the temperature increases, rather than decreases, with altitude. In regions of inversion, rising air masses lose their buoyancy, so water condensation and rain are suppressed. Smoke could be stabilized and would persist in the dry upper atmosphere, shielding the earth's surface from sunlight. That situation could continue for long periods, and the resulting temperature change would damage all life in the underlying region. The argument, as continued by Sagan, holds that smoke may well be distributed over the entire globe and that the survival of the human race could be endangered. Although the findings of Turco *et al.* are inconclusive, Sagan calls them scientifically robust.[19]

Examination of the possibility of nuclear winter differs from ascertaining fallout and ozone layer depletion, because it depends on meteorological phenomena that involve much more detailed and complex considerations and calculations. Several important meteorological processes are inadequately understood, and the bases for smoke estimation are very uncertain. Computer modeling sufficiently detailed and refined to handle all the various pertinent factors is not yet available.

It would take about two weeks for the smoke to disperse over wide areas. Turco *et al.* do not consider in sufficient detail the mechanisms that may be operating during that period, which could sharply decrease the amount of smoke available.[20] Weather predictions for five days in the future have some validity. Calculations of meteorological processes two weeks in the future have not yet been successful.

The average tropospheric residence time of fine dust particles, such as those containing radioactivity, varies between one and three weeks.[21] The average residence time of water vapor in the atmosphere is little longer than a week.[22] Studies of the continuing discharge rates of man-made and natural smoke, and observations of the average amount of smoke found in the atmosphere, suggest that smoke may have an atmospheric residence time of a week or less.[23]

Under all atmospheric conditions, water is the agent largely responsible for cleansing the atmosphere of smoke. Turco *et al.* estimate that 225 million tons of smoke particles would be emitted in their 5,000-megaton baseline war.[24] A comparison between the weight of smoke and the weight of the water (droplets and vapor) in the northern latitudes between 30° and 70° shows that the water outweighs the smoke by a factor of ten thousand. The average lifetime of a water molecule in the atmosphere is little more than one week. So, in ten days, before the widespread smoke-associated cooling effect is established, a mass of water a few thousand times greater than the total mass of war-engendered smoke will rain out of the atmosphere. Turco *et al.* estimate that a factor of two or three of the smoke will be removed by rainout in the fourteen-day period. That factor is critical and quite uncertain.

The highly relevant question of how much smoke will remain aloft after two weeks of rainout is unanswered, because it depends on the relative altitudes of water and smoke, on alterations in scavenging processes, on latitudinal variations, and on various other factors. Some studies suggest specific cleansing factors.

The capping clouds that regularly appear in the smoke plumes above big fires are produced by condensed moisture that forms droplets too small to develop into rain. In one observation made for a U.S. Environmental Protection Agency 1983 report, L. F. Radke and colleagues measured the particle size distribution for smoke following two different paths, one through the capping cloud and one below it.[25] The cloud reduced by a factor of 10 the number of small smoke particles between one-tenth and one-half micron in size. The effect would be large because most of the particles would pass through the cloud.

Changes produced by the capping cloud may result from each of the water droplets trapping many small smoke particles. Upon subsequent evaporation of the water droplets, those particles are agglomerated into bigger units. In addition, Hanel observed that aerosols increase in size if they are exposed to air in which the relative humidity exceeds 90 percent.[26] Increasing the sizes of the particles reduces their number in the range where they are most effective in absorbing and scattering the sunlight. Larger smoke particles rain out and fall more rapidly. Thus, at the very origin of smoke generation the particles already may be prepared for more rapid deposition.

The Turco *et al.* model for predicting nuclear winter did not include the effects of oceans.[27] The nuclear winter mechanism for the abrupt cooling of the earth's surface and the lower troposphere is effective only over land. Over the oceans, the surface water is continually supplied with heat from below. For periods of many months, the temperature at the ocean surface would remain essentially unchanged.

The National Center for Atmospheric Research (NCAR) uses a somewhat more realistic model of the atmosphere in that it includes oceans.[28] In the NCAR model, the extent of the temperature reduction is lower by a factor of two to ten (depending on the season) than in the Turco *et al.* study. The cooling is greater in summer and much smaller in winter.

However, the NCAR model still postulates that the injected smoke layer appears instantaneously at all temperate zone longitudes. Furthermore, the NCAR model postulates that solar radiation is absorbed by the smoke layer as though the layer were fixed in its longitudinal position and concentration. The postulate ignores the motion of the smoke produced by convection processes related to natural air flow, although other consequences of the wind are taken into account. Finally, in the absence of a usable model, the NCAR study postponed a consideration of the most important factor, rain-out.

If a nuclear winter were to begin, great temperature differences would be established between continents and oceans. That would result in storms along the coasts of the continents, with consequent mixing of air from high and low levels. Air from lower altitudes has a higher moisture content, which is conducive to rainout. If the rain removed smoke from one area while leaving it untouched in another, the smoke layer would be patchy. The obstruction of light is most effective when smoke is well spread. The hemispheric light reduction would decrease as patchiness increased. The patchiness of smoke, in turn, could bring about further large temperature differences, which would further drive the atmospheric processes that rain out more smoke. In that case, rainout of large smoke layers might proceed in an accelerating fashion.

Some smoke particles, particularly those from oil fires, absorb light strongly. At first those particles may be water-repellent, as Turco *et al.* suggest.[29] That property changes after relatively brief exposure (twenty-four hours) to natural oxidizing agents in the atmosphere. Therefore, there is little reason to expect that rainout of all types of smoke particles would not occur.

In conclusion, there is no obvious reason why large quantities of smoke particles should have a long residence time in the atmosphere. The calculations associated with the nuclear winter do not include major cleansing effects of water vapor, which are themselves smoke-induced, or the influence of the oceans and winds during the time needed for spreading the smoke worldwide.

A recent detailed study by A. A. Broyles, based on observations of forest fires and considering the proper modifications pertinent to a discussion of burning cities, proposes that the upper limit of smoke that might be emitted in a nuclear war is 360 million tons, while the lower limit is 15 million tons.[30] The Turco *et al.* estimate for their baseline war producing 225 million tons is in the upper range.

The uncertainties associated with the amount of smoke that would be emitted are extremely numerous. The ratio of wood burned to smoke emitted depends on many factors, including moisture content, the temperature of the air surrounding the fire, the manner in which the wood is packed, and the accessibility of oxygen. The last two factors depend respectively on the size of the object burning and on the wind velocity.

Estimates of the quantity of smoke generated are based on assumptions about the weight of the smoke emitted relative to the weight of the fuel burned. The measured smoke emission factor presented in one forestry report ranges from 0.0025 to 0.063 for wood.[31] Measurements made at the Georgia Technical Institute give values ranging from less than 0.01 to 0.2 for plastics.[32] There is some possibility that large urban fires may provide sufficient heating of burning components to consume even more of the smoke than is indicated by those numbers, so that low values are applicable.

City fires produce more smoke than forest fires, because the fuel density in cities is estimated to be between 20 and 200 kilograms per square meter (kg/m^2) but only between 0.5 and 10 kg/m^2 in forests.[33] In cities with the higher fuel density values, fire storms are possible. Fire storms that loft smoke to very high altitudes are relatively rare events, apparently dependent on the presence of dense fuel and particular meteorological conditions under which practically all available oxygen is consumed. The possibility that fire storms may loft smoke into the stratosphere cannot be excluded but has never been observed. Turco *et al.* assume that only 5 percent of the smoke produced is due to fire storms.[34]

Smoke generated by burning petroleum products, rubber, and plastics tends to absorb light more strongly than the smoke from burning wood. The predominant fuel in cities, at least 95 percent, seems to be wood and paper products.[35] The strength of light absorption of city smoke is critically dependent on the proportion of petroleum products, plastics, and rubber present. Petroleum refineries could provide a total of 150 million tons (1.5×10^{14} grams) of fuel in the United States.[36] The weight of smoke would be about 15 million tons, which is a small fraction of the smoke estimated. Oil found or stored underground is not apt to be ignited. Thus, the strongly absorbing carbon-containing smoke from that source is small in quantity.

An obvious uncertainty is connected with our inability to predict how a war would be conducted. If only purely military targets were attacked with relatively small bombs, the amount of smoke generated would be much lower. Because of the trend toward more accurate, lower-yield missiles, indiscriminate damage, the source of great quantities of smoke, is far less likely. The deployment of anti-ballistic missles to defend cities (as in the case of Moscow) would further reduce fire and smoke.[37]

CONCLUSIONS

During the two-week period required to establish the full effects of a nuclear winter, the water in the atmosphere, ten thousand times greater in weight than the postulated emitted smoke, most probably will rain out. That alone could reduce the smoke content at low altitudes many times. Other weather phenomena, enhanced by the patchiness of the smoke and the temperature differences between land and ocean, will serve to bring smoke from the upper levels of the troposphere to the lower, whence it can be removed. The immediate coagulation effect of capping clouds on the smoke has been observed to reduce the small smoke particles emitted (which are the most dangerous) approximately tenfold. The targets that would be attacked are unknown. The proportions of highly light-absorbent smoke that would be emitted are also unknown. Complex factors such as fuel density, moisture content, air temperature, and other properties related to smoke emission make current estimates unreliable. The associated effects of each of those uncertainties on temperature change tend to decrease the amount of smoke retained in the atmosphere. The effects of the uncertainties are large enough that revising a few of the smoke estimates and including a few meteorological effects invalidates the prediction of a severe nuclear winter.

Given the uncertainties and omissions in the theory on which nuclear winter is based, the concept of a severe climatic change must be considered dubious rather than robust. Nonetheless, the *possibility* of nuclear winter has not been excluded.

A moderate decrease in temperature is more likely than the probability of the occurrence of extreme temperature changes. It is quite probable that over widespread areas the temperature aloft will become higher than the temperature on the ground. Such a temperature inversion decreases rainfall. Nuclear winter seems unlikely, but widespread crop failure and famine may well follow a large-scale nuclear war.

There is every reason to undertake the difficult task of arriving at more realistic predictions. More funds should be allocated for experiments to sample fire plumes and to determine how emissions scale with the size and type of fire. Greater amounts of money, carefully spent on atmospheric modeling and experiments, would accelerate resolution of the basic questions regarding climatic changes.

From biblical times to the present, written records testify to times of food shortage from causes ranging from volcanic activity to mankind's folly. The amount of suffering that would be produced today by the failure of a single year's harvest is horrifying to consider. Extensive food storage could decrease greatly the risk of famine and could be easily accomplished in the United States and in many other nations. The usefulness of such storage is unquestionable.

Speculative theories of worldwide destruction, even the end of life on earth, used as a call for a particular kind of political action, serve neither the good reputation of science nor dispassionate political thought. Informing the general public about the reliability, unreliability, or incompleteness of such studies seems an important responsibility, for only then can decisions affecting the well-being of society be made on an intelligent basis.

*   *   *

Author's postscript. There is considerable but unproved evidence that a decrease in sunlight may have caused the extinction of dinosaurs and many other species at the end of the Cretaceous period. Alvarez and Alvarez first suggested that the presence in various sites throughout the world of an iridium layer deposited 65 million years ago is evidence that an asteroid hit the earth at that time.[38]

Found in small concentrations throughout the universe (particularly small in the crust of the earth), iridium is a most inactive element—more noble than platinum or gold, which are similar metals. Its lack of reactivity has led to its persistence at the place it was deposited. The amount of iridium in the layer suggests that the meteorite must have been five to ten miles in diameter and its crater between seventy and one-hundred miles wide. An impact by so huge a meteorite is by no means impossible. There is some evidence that a huge meteor crater exists in the Indian Ocean north and somewhat east of Madagascar. The Bering Sea has also been suggested as the site of the crater.

The hypothesis suggests that the meteorite penetrated the atmosphere, crashed into the surface of the earth, and that the energy of impact created a fireball so large that it reached the stratosphere, carrying a huge quantity of material from the meteor and its crater. The theory postulates that the material condensed there and excluded sunlight, which in turn affected plant and animal life so severely that many species, particularly large animals, became extinct. Subsequently the material, which contained iridium, was deposited in a rather uniform layer on the surface of the earth.

Some clay samples from sites of the same geological period were shown to contain, in addition to the iridium, a significant amount of soot.[39] The investigators suggest that the fireball formed by the collision killed much vegetation and set forest fires. The soot from those fires, carried to the stratosphere, darkened the sky and contributed to the consequent biological catastrophe. The estimates of the amount of energy released and the amount of material distributed around the earth are approximately one thousand times greater than the amounts of smoke postulated by Turco *et al.* as the cause of nuclear winter.

The meteor theory remains controversial. There is no proof that a huge meteor struck the earth. Saurians may have died out for different causes, or the meteor may have been only an added cause. However, further confirmation of the hypothesis would seem to provide evidence that life persists through catastrophes even greater than nuclear war.

To diminish the likelihood of the catastrophic impact of warfare, one should give most serious consideration to defensive measures. Such measures may not only deter war but, in the horrible case of actual hostilities, may also substantially reduce both the local and widespread effects of a nuclear war.

18

A Note About Civilian Protection

For more than two decades U.S. policy demanded that the ability to deter nuclear war be based solely on retaliatory strength. All forms of defense, including passive defense of civilians, received little or no attention. During that same period four other nations—Switzerland, Sweden, China and the Soviet Union—developed and deployed extensive civil defense measures.

With the initiation of the SDI program in 1983, the role of defense measures in decreasing the probability of war was at last brought to public attention. Deterring war is the most important benefit that defensive measures—or any other measures—can provide. That point has been lost in the debate over what percent of the attacking missiles a defensive system could destroy. Actually, the danger of a first strike—which in order to be successful must wipe out all possibility of effective retaliation—is decreased by every defensive measure that is deployed.

Complete safety and security were not possible in the past; they are not going to be available in the future. What we can achieve is an improved ability to deter war. The beneficial effects of civil defense in a combatant country will be limited, but in conjunction with strategic defense, some programs can be very important both as a deterrent and in reducing suffering and deaths.

In the United States the Federal Emergency Management Agency (FEMA) is responsible for civil defense together with all other forms of protection against natural disasters. Evacuation procedures are an important part of the FEMA program. In November 1986 the director of FEMA Lieutenant General Julius W. Becton, Jr. spoke to a conference of private citizens interested in supporting civil defense. He began by telling an apocryphal story about a minister whose church was threatened by a flood.

The minister, it seems, turned back the first evacuation workers who came by car to take him to safety. "The Lord will save me," he explained. The flood waters rose, and a rescue crew approached the church in a rowboat. The minister again refused to leave: "The Lord will take care of me." Finally, the water drove the minister to the roof of the church. A helicopter approached and offered to lift the minister to safety. He steadfastly remained: "I trust the Lord." The waters rose further, and the minister drowned. At the Pearly Gates, the minister complained to Saint Peter: "Why wasn't my faith rewarded?" Saint Peter answered, "We tried three times. What did you expect? A miracle?"

General Becton's talk was very clear, very matter of fact. He expects a little more support for civilian protection but not a miracle. I have realized, and I hope others will realize that moderate support for civil defense may make a great difference. One possible means that has not received sufficient emphasis is evacuation as a response to a threatened attack.

The Soviet civil defense program is more than three decades old. It includes secure shelters for party and government officials and less elaborate shelters for some workers in important industries. The Soviets have stockpiles of essential materials, evacuation plans for nonessential city dwellers, and designs for easily improvised protective shelters for evacuees.

The effectiveness of some of those preparations was demonstrated in the recent Chernobyl accident. The Soviets reported that the people of Pripyat, the town nearest the reactor, were notified of the accident and necessary precautions taken during the early morning hours after the accident. Tablets to minimize the uptake of radioactive iodine were distributed the same day in nurseries and schools. The evacuation of 45,000 people from Pripyat on the following day was accomplished in three hours time. FEMA would be hard pressed to match that performance.

In the United States many people seem to doubt the need for and the efficacy of evacuation procedures. Evacuation seems impossible because less than a half hour would remain after an actual attack began. That ignores an important fact: A complete surprise attack is a most unlikely beginning for a nuclear war.

Threats of war are less horrible and much more likely than a war that begins with twenty minutes notice. One possible situation is that, in a crisis, the Soviet Union might threaten war in order to achieve its ends. Soviet blackmail to be credible would have to include preparation for a nuclear strike, including Soviet urban evacuation. If the President had the ability to safeguard a high percentage of the population by counterevacuation, the possibility of such a threat would disappear. The solution is not novel. It is as simple as taking out insurance.

Opponents of civil defense argue that the Soviet Union would never threaten war under any circumstances. They may be right. But the rapid buildup of Soviet weapons of mass destruction during a period of voluntary decrease in American nuclear weapons cannot be ignored. Today the Soviet Union possesses more than half of all the nuclear missiles in the world. The multiple and extensive Soviet defense measures and their present propaganda against American defense are added reasons for establishing at least some protective measures.

The United States is the most mobile country in the world. With planning, if a Soviet evacuation were observed, the President would inform the public of the fact and announce the beginning of voluntary evacuation. People would be instructed by radio and television as well as by recorded telephone messages and posted bulletins about what to take, where to go, and when to leave.

The traffic jams that occur on holiday weekends, when a considerable portion of the city residents travel, may make such a plan seem impractical. However, in spite of slow traffic movement during these peak hours, major exodus and return are completed—even without planning—in comparatively short time spans. There is ample reason to believe that preplanned evacuation can be conducted effectively.

In the United States evacuation is and will probably always be voluntary. That means a little planning could make a great difference. Advance planning must provide for dispensing information, determining and scheduling evacuation routes, organizing transportation for those without private cars, and preparing reception areas. No one should be forced to evacuate. Those preferring to go to an area other than the one suggested (because of friends or relatives in a different outlying area) should be free to do so. The strength of our society lies in voluntary choices. Utilizing that characteristic as far as possible during a critical period could be of great advantage.

Evacuation traffic patterns would include a departure schedule for the various areas of each city. Major streets would be converted into one-way arteries leading out of the city, and access to them would be scheduled for various city regions. At important nodal points, traffic policemen would direct traffic, and tow trucks would be available to keep those critical areas clear. Traffic police would notify evacuation centers of stranded motorists. The transport vehicles from evacuation centers would then stop along the evacuation route during their regularly scheduled trips to pick up those stranded.

Food and medical supplies sufficient for at least two weeks must be stored in the reception areas. Safe storage places for these supplies could be made available at low cost. Food supplies could be more easily provided in the United States than in almost any other country. For example, the amount of dried milk already purchased by the government, if stored in reception areas, would suffice for many months. Grains currently owned and stored by the government could also be dispersed. While such preparation would involve some expense, careful planning and, possibly, tax incentives to large food wholesalers for establishing safe storage centers in outlying areas could minimize costs.

The important and inexpensive parts of civil protection are planning the evacuation and the provisioning of the areas that will receive the city population. The expensive part is to provide the necessary shelters for the people who must remain behind: policemen, firemen, and personnel to operate water, gas, electric, and communication services. Expense can be limited by keeping the number of such personnel as small as possible. Most facilities should cease operation during the evacuation period. Wherever feasible, production should be replaced by stored goods.

Shelters will also be needed for those unwilling or unable to leave. That is, of course, the real point of civil defense. The Swiss government has spent great amounts of money to shelter all its citizens from the dispersed effects of a nuclear war. Over the last thirty years in a quiet way, the Soviet Union has spent tens of billions of dollars on shelters. It is obvious that our own defenses must remain incomplete. No one can say and, indeed, no one should say that a nuclear war would be anything other than a terrible disaster. Civil defense in general and a plan to evacuate in particular serve primarily the purpose of making such a disaster less likely.

The beneficial effects of civil defense will be limited, but, in conjunction with strategic defense, the combined effect can be important. Among the actual consequences of nuclear war, the burning of cities is the most horrible. Defense of cities is indeed the most important part of strategic defense. We cannot ever know how well such a defense will work. But trying to accomplish defense remains a necessity. A combination of passive and active defense will make the prospect of a successful attack less certain. This is the essence of the argument for any kind of defense.

A plan for counterevacuation is a particularly fitting complement to SDI. Strategic defense requires the development of high technology; evacuation is based on a strength present today. FEMA need not change its nature. Evacuation is currently used when a natural disaster threatens. It can also protect against the great unnatural disaster of war.

Evacuation of urban populations is one of several ways in which the disastrous consequences of war could be limited—at least to some extent. This, together with other civil defense alternatives, should be considered. Evacuation, of course, would depend in a critical way on the cooperation of the great majority of the people.

New York City has suffered two blackouts during the last few decades, the second of which gave rise to extensive looting. I was in New York during the first blackout. New Yorkers on that occasion displayed a remarkable spirit of cooperation—the people on the streets became friendly and helpful. I have never enjoyed New York as much as I did that night. In small troubles as well as in the greatest difficulties, people's reactions count. No plan will work without cooperation. But without some planning, no matter how positive the people's reactions may be, disaster cannot be avoided.

A usual and easy response to a distant danger is to forget about it. But because we have not quite forgotten, the danger visits us in our dreams. We must think more about concrete ways to face the danger of a nuclear war. Active discussion of civil defense is neither comfortable nor popular. In our time of uneasy peace we must use foresight and imagination to preserve peace. The particular solution that I recommend here seems to me not quite as important as the need that we consider every possibility to preserve human life.

─── IV ───

PROGRESS
AND
PARADOX

19

Technology in the Developing World

*   *   *

Author's Note. The following essay is an adaptation of a lecture I gave to a group of students, faculty, and representatives from various Indonesia government agencies at the Bandung Institute of Technology in October 1975. My host Professor M. T. Zen was particularly helpful in acquainting me with the concerns and problems of his nation, and the Mitre Corporation generously sponsored the trip and recorded and published the lectures.

The world has become small. What happens anywhere is known everywhere; it has repercussions on all continents. Technology moves much faster than people can change their minds. Irrespective of the outcome, change itself is painful. That may be why many young Americans criticize the age of the machine and want to turn back to nature. To appreciate technology, we must recall the history of ancient times. We remember the splendor of ancient Greece; yet the Greek society was based on slavery. New ideas, beautiful sculptures, anything of lasting value was created by the few at the top of the pyramid; the majority had to be slaves. That was considered the immutable nature of human society. Technology has opened the possibility of freedom for everyone.

I am not saying that technology is all good. I am saying that technology is a challenge. Whether it is mostly good or mostly evil depends upon us. The hope that technology would abolish both slavery and poverty has not been fulfilled. The manner in which technology is entering the developing world is related to those two ancient scourges.

Most happily, freedom has arrived in Indonesia, thanks to the determination of the Indonesian people. But freedom does not mean happiness or the absence of mistakes. Freedom means that you are free to make your own mistakes. My own mistakes are better mistakes: I can correct them. If the mistakes belong to somebody above me, I cannot correct them. I react with hatred, and hatred is destructive. Freedom means responsibility and the possibility of happiness.

Was freedom achieved throughout the world when colonialism—an old form of suppression—crumbled? Do not be certain that colonialism is gone. I come from Hungary. In Eastern Europe a population about the size of the population of Indonesia lives under Russian rule. Some of those people used to rule themselves. I cannot forget it; I hope others will not forget it either.

About the year 1200, Genghis Khan conquered a large part of the known world. In Russia and in China, where he established empires, a lack of freedom has continued to this day. The current government is called communism. It is not new. Its roots go back almost eight centuries. It has little to do with Karl Marx. It has a lot to do with old traditions in Russia and China, with the tremendous difference between the rulers and the ruled.

Still, a great part of the world today is free, for better or for worse. In a large part of the world, poverty is a problem. The solution does not seem even to be on the horizon. As far as economic conditions are concerned, freedom has not yet worked out for the better.

In talking about a solution, I am going to commit the crimes of oversimplification and exaggeration. If I could not do so, I could not talk. One of my fondest memories is a discussion I had with the great physicist Niels Bohr when I was a young man. The subject was a detail that I had studied, and I contradicted Bohr. In replying, Bohr said, "Teller knows a hundred times more about this than I do." I impolitely interrupted at that point and said, "This is an exaggeration." Bohr glared at me as though he were angry and said, "Teller does not want me to exaggerate. If I can't exaggerate, I can't talk. All right, I agree with you; you only know ninety-nine times as much as I do." There was no possible reply to that. Now I shall exaggerate; so did Bohr.

There are two ways to deal with poverty. One is to make a firm overall plan to change the social structure. Some social structures need to be changed, and it is very tempting to adopt a plan and, like the Soviet Union, implement it ruthlessly. The plan may help poverty, but its price must be paid in terms of freedom.

Another way is to use technology. Instead of planning to divide what is, try to create more. That is not an easy program, but it has great advantages. It is compatible with freedom, and it does not lead to conflict between people. The most wasteful way to use human energy and human intelligence is to use them in conflicts.

What everyone feels most directly is not how well he lives but whether he lives better than yesterday. Some may consider it unjust if half the world is rich and half the world is poor. It is even more unjust if one-fifth of the world is rich and the rest is terribly poor. But trying to correct such a situation by redistribution will cause fights, and in a fight everyone suffers. What is needed is to create more.

How to do so is often misunderstood. Communists and capitalists agree that capital is needed to create wealth. I believe that statement of the situation is a dangerous half-truth. How important is capital? Japan and Western Europe recovered from World War II at a remarkable rate. The underdeveloped world has made slow and uneven progress. Why? The question is more important than the answer. I may be wrong in my answer; I am certainly not wrong in raising the question.

The people in Europe and Japan knew about technology in all its aspects. They knew how steel behaves, how to maintain a machine, and, last but certainly not least, how to cooperate in an industrial society. That human element may be called knowledge of management, but it is not confined to the managers; it is also shared by those who are managed. Because of that knowledge, people accept an organizational structure in which they can be free, can contribute, and are rewarded. That knowledge helped Western Europe and Japan to recover in a way that was called an economic miracle. What is not expected by economists is invariably called an economic miracle.

A limited American contribution helped the recovery. In every technical development there are bottlenecks—essential ingredients that happen not to be available. The important role of the American aid was to break the bottlenecks. But the main drive came from within Western Europe and Japan. Capital formation was not the cause—it was the consequence of the technical development.

In Western Europe and Japan, knowledge, know-how, was available but capital was missing. The result was miraculous success. Is there a case where capital is available and know-how is not? There are some striking examples of that situation. One is Saudi Arabia. Never has there been as much capital as quickly accumulated as on the Arab peninsula. General education was lacking. Some of the princes have learned about economics at Harvard. They have not learned about technology. In Germany and Japan, a little capital coupled with widespread knowledge resulted in success. A superfluity of capital and little common knowledge have not produced notable general improvement in Saudi Arabia.

A tragic variation of this situation—considerable capital coupled with some effort toward social reform—has been seen in Iran. The Shah followed his father's efforts toward technical improvement by turning billions of petrodollars into new industrial plants. Neither educated labor nor the necessary management was available. A genuine land reform was not followed by developing agricultural knowhow in the farm population. Unprepared millions were attracted to the seemingly rich city of Teheran. The population of Teheran increased almost tenfold; it became a city of 5 million with no sewers. Simple people with insufficient knowledge turned to a bigoted religious dictatorship. The few able experts, carriers of technical understanding, fled their homeland.

The story of the development of Soviet technology is different. A widespread tradition of technology was absent, but a clearly directed, strong drive toward technological knowledge had already begun before the war. Children in Soviet schools knew that their lives would be miserable and insecure unless they learned, and science and technology were on the top of the list of subjects. A little more slowly than Western Europe and Japan, but in a most impressive way, the Soviet Union not only repaired the ravages of the Nazi invasion but also developed a substantial economy.

Soviet technology has its strong and weak spots. The Soviets have done excellently in developing their military forces, have done well in developing their heavy industry, and have done a poor job of providing consumer goods. They have made a shambles of the agricultural sector.

Why the disparity of success? The soil and climate in the Ukraine are good. Some Soviet colonies, like Hungary, used to export food. Now, even though the population has not grown, Hungary has to import food. Part of the answer lies in Marxist doctrine, which holds that countries that produce raw materials are exploited by rich industrial countries, which produce manufactured goods. Progress and wealth, according to that view, are dependent on industry. Backwardness is equated with raw materials. The result is that talented and ambitious people in communist countries ignore agriculture.

That is too simple a picture. Stalin's forcible collectivization—the man-made famine of the 1930s that killed millions of Ukrainian kulaks—must be mentioned. The bizarre case of Trofim D. Lysenko—who taught that the laws of inheritance are not valid, that ordinary wheat seed would adapt itself quickly to an arctic climate—was added to the collectivization disaster. Stalin chose to believe the charlatan Lysenko. The able geneticists who disagreed with him were sent to concentration camps. The wheat and the geneticists died. It takes a long time to grow new geneticists.

What happened in the Soviet Union also fits my oversimplified thesis. Marx did not know the importance of agriculture, and in the Soviet Union Marx cannot be contradicted. Technical understanding among the people is decisive whether one discusses industry or agriculture.

No one considers the United States an underdeveloped country, but the United States today imports more manufactured goods than it exports. The country would go broke except for one fact: About one-half of the corn and the grain that appears on the international market comes from the United States.

That success began in 1862. In the middle of the terrible Civil War, Congress established land grant colleges devoted to the agricultural and mechanical arts. The initials in Texas A&M stand for agriculture and mechanics. Before that time, the United States was growing rapidly but was still primitive. The Land Grant College Act transformed the United States from a backward country, a borrower of money, into the most wealthy and, at least for a time, the most powerful country in the world. The reason was technology applied to the mechanical arts and to agriculture.

The United States is the greatest producer of rice and soy beans. Cotton is another important U.S. agricultural product. While cotton cannot be eaten, cottonseed is nutritious for animals with more than one stomach. A chemical in cotton seed makes it poisonous for humans, but a cheap new technical procedure can extract the poisonous ingredient. The new technology could be important for India.

Whatever else you believe or disbelieve about Marxist theory, do not believe that agriculture should be neglected. The biggest problem facing Indonesia is to improve agriculture and its products by new procedures—irrigation, fertilizers, systems of distribution, conservation of food. All those are part of a technological structure. Southeast Asia could produce twice as much rice on the same acreage as it is producing now. Technology could provide the means to increase agricultural production. Include your outer islands, not necessarily for growing rice, but for growing crops appropriate to those regions.

One possibility may be especially promising for Indonesia. As a nation of islands, the real estate covered by water is more extensive than the real estate above sea level. Today a small percentage of your food comes from the sea. It could be a big percentage, particularly if you could abandon the Stone Age method of harvesting without sowing, of catching fish without first cultivating fish. In China, fresh water fish have been cultivated for more than a thousand years. The ocean is a magnificent potential resource. Fresh water fish and brackish water fish have been cultivated. Salt water fish are just beginning to be studied. We know little more about their diseases and survival than we knew about medicine before Pasteur. Cultivating salt water fish would be one way for Indonesia to feed 200 million people and more.

Almost two centuries ago Thomas Malthus said that the population explosion would result in catastrophe. The fertility of mankind is great indeed, but the inventiveness of mankind is greater. At the time Malthus was writing, no country in the world, including his own England, could have been called a developed nation. Yet in spite of the population increase, the standard of living has improved greatly. Technology has grown faster than population.

Technology does not mean steel mills. It does not even start with steel mills. It may start with rice paddies, or it may start on the shores of the Java Sea. Throughout the world people will breed fish first along the shore and then proceed to ranching fish—breeding the young, allowing them to swim out into the ocean. One also should weed the oceans and keep down the number of predators. There can be no agriculture if weeds are allowed to proliferate. Ocean cultivation is difficult, but it could feed many more people than the present world population. Today, 7 percent of all food comes from the oceans, but 20 percent of the high-grade protein available is furnished by this source.

Oil, manganese nodules, and probably many other useful materials are present on the bottom of the ocean. You have your own private sea, part of it shallow, part of it very deep. In this instance, you have the opportunity to leap ahead, and if you do, we in America will be happy to come and learn from you.

Indonesia is discussing the possibility of putting up a satellite. A stationary satellite is possible only on the equator, and the equator is part of one of Indonesia's territorial possessions. I suggest that the satellite interact with a large area extending beyond Indonesia. A wider weather map allows for better weather prediction; winds know no national boundaries. Put the best, most modern observation equipment into your satellite. Observing clouds, following the winds, measuring the temperatures at different altitudes make it possible to predict the weather. If you predict the weather, you will be in a better position to save your harvest.

Here in Indonesia, population growth is creating problems. Unlike the participants in the Club of Rome, I believe that the world's supplies will not be exhausted and that the changes produced by technology need not be evil. We can reshape the world into a form closer to our hearts' desire. I believe that the ultimate limit to the population expansion is our ability to get along with each other. The ultimate problem is how man can tolerate man. The ultimate problem is peace. If we want to avoid war, we need understanding. We must understand nature and technology. But the most necessary and difficult thing is to understand ourselves and each other. Can people ever be understood?

Your satellite should be called a communications satellite. The job of such a satellite is to communicate, to educate, and, perhaps, to do something more. American technology has created the marvels of television. Yet no one has captured on television the image of our world and the accomplishments of the human mind. We have found the means of communication. We have not found the substance that we should communicate.

My good friend, Dr. Zen, who suggested that I lecture in Bandung, asked me about technology: What should Indonesia do? Should we copy? Should we leapfrog? About communication I can offer an answer. There is nothing worth copying. If you leapfrog in communication, you will have developed not only a new application but also a new meaning for technology. You will have created the art of the new age.

20

Chernobyl and the Safety
of Nuclear Reactors

The September 1986 issue of the Soviet English language magazine, *Soviet Life,* contains an article on the Chernobyl accident. An enlarged quotation on the second page reports:

> Blind chance sweeps the world along. Of the 41 operational reactors in the Soviet Union, chance chose the newest, which went into service in 1983. (The USSR put the world's first reactor into operation in 1954.)[1]

The explanation of the most severe reactor accident in history is necessarily complex. Understanding the events and their causes requires patience and an open mind. There is more reason to talk about human blindness, however, than about blind chance.

In the West the first sign of trouble came on Monday, April 28, 1986, when radiation detectors in Sweden showed several hundred times the usual level of activity. Possible sources of radioactivity in that nation were immediately checked. Nothing out of the ordinary was found. The wind was coming from the Belorussia-Ukraine region. At first the Soviets denied any knowledge of an increase in radioactivity. The news from Chernobyl at 1:40 A.M. on Saturday, April 26, seems to have been received in Moscow with incredulity—everyone knows that a nuclear reactor cannot blow up. A nearby reactor continued to operate for three and a half hours while firemen were battling fires in and around its building. Permission to shut down the other reactors in the complex was not given until Sunday. Not until the end of the first working day following the accident did the Soviets announce that a major nuclear reactor accident had actually occurred.

In a more extended statement in mid-May, General Secretary Gorbachev said: "It is hard to imagine the future of the world economy without nuclear energy.'' The comprehensive report issued in August makes it clear the Soviet Union will increase its use of nuclear energy fivefold to sevenfold by the end of the century. But in the free world where public opinion determines policy, the psychological fallout of the Chernobyl accident may discourage extensive use of nuclear energy.

That would have serious consequences. The world's supply of oil is limited. Oil at present is irreplaceable as a source of energy for developing nations because it is easily transported and used by the machinery available in those regions. Continued extensive use of oil by developed nations results in higher costs per barrel and is likely to condemn the Third World to long-term poverty. Of all the forms of pollution extant today, pollution by poverty has the most severe effects, both in loss of life and in the destruction of the environment.

Sweden, where 42 percent of the electricity is generated by nuclear reactors, was the first nation to experience and identify the radioactivity from Chernobyl. In 1980 the Swedes voted 3-to-2 to convert to nuclear power. Their planning enabled them to establish nuclear plants at a reasonable cost, together with a safe system of waste disposal. Today, the lower oil prices, somewhat more secure sources of oil, and the fear of radioactivity perpetrated by sensationalized media reports may make nuclear energy appear less desirable. Yet most of the world's oil reserves are in the Middle East, a politically unstable region. Unnecessary reliance on oil by the free world will not only raise the price of energy and harm the developing world but will also invite political blackmail and threaten peace.

In the United States, work on reactor safeguards began more than thirty years ago. In 1948 the Atomic Energy Commission (AEC) was established. One of its missions was to oversee the development of nuclear energy for generating electricity and for research. The AEC set up an Advisory Committee on Reactor Safeguards to evaluate the safety of reactors. I was the first chairman of that committee.

The first item on our agenda was to determine if, under any conditions, a nuclear reactor could explode like an atomic bomb. The finding was that it could not. Both nuclear explosives and reactors depend upon a multiplication of neutrons. But a nuclear explosion requires that no reaction occur until the whole mass is assembled. If there are neutrons present beforehand, the fissionable mass produces heat and radiation but no explosion. A reactor is constructed to ensure a constant supply of neutrons, the opposite of the condition needed for a nuclear explosion. There is no way the supply of neutrons can be excluded from a reactor. Nevertheless, our committee found that very serious problems, including fires and conventional explosions, could arise in several ways.

Fission occurs when a neutron strikes an appropriate nucleus, which splits and releases two or more fast neutrons. Those neutrons will escape, will be absorbed in a non-fissionable material, or will cause fission. A reactor is *critical* if the neutrons produced by fission are equal in number to the original neutrons that initiated the fission.

However only 99 percent of the neutrons produced by fission are released promptly; about 1 percent are released after a delay of about one minute. If some delayed neutrons are needed to make the reactor critical, the power will not change much in a minute, which is sufficient time to adjust the control rods. In that case, reactors can be operated at a steady rate of power without much difficulty.

If the total of prompt and delayed neutrons produced by fission exceeds the original set, the reactivity rises slowly; the percentage of excess is conventionally described in *cents*. At *10 cents,* the arrival of 90 percent of the delayed neutrons will replace the original neutrons. At *50 cents,* 50 percent of the delayed neutrons will bring the total up to critical. At *100 cents* or *one dollar,* the prompt neutrons alone replace the original number of neutrons, and the reactor is *promptly critical.* This state is very difficult to bring under control. But a worse case exists. If neutron-absorbing material is lost from the reactor, a great number of prompt neutrons, more than the original set, may become available to produce fission. The reactor is then likely to become uncontrollable, a *runaway.* The reactor does not explode like an atomic bomb, but great heat can build up, produce fires or conventional explosions, and release accumulated radioactivities.

Beginning in the late 1940s, our safeguard committee based its decisions on the assumption that all serious accidents must be avoided. A single serious accident might block all further development of nuclear energy. The main potential dangers are related to five effects. One was discovered before the AEC existed. During World War II, the United States built three large graphite reactors at Hanford, Washington. Neither the reactors nor the weapon whose material they supplied had been imagined ten years earlier. Yet both worked at their first trials. However, a surprise occurred the first time a Hanford reactor was shut down: It could not be started again until a day passed. Enrico Fermi provided the explanation, which is relevant to Chernobyl.

One of the many fission products present in a reactor is iodine-135, an unstable isotope that decays (with a halflife of 6.7 hours) into xenon-135. Xenon-135 has one very unusual property—it is a million times more effective in capturing neutrons than most neutron absorbers. In an operating reactor of commercial size, xenon-135 quickly acquires a neutron and becomes xenon-136. However, when the reactor operates at very low power or is shut down, radioactive iodine-135 continues to decay into xenon-135. The accumulation of xenon-135, called *xenon poisoning,* then prevents the chain reaction from proceeding until it decays (with a halflife of about nine hours) to cesium-135. The Hanford reactor would not start after shutdown because of xenon poisoning, which in an operating reactor can give rise to complicated and unexpected situations.

A second potential source of difficulty was also recognized during World War II, a problem called *Wigneritis* after my friend, Eugene Wigner, who first explained it. In most reactors, the fast neutrons released by fission have to be slowed or *moderated* to continue the chain reaction efficiently. In graphite-moderated reactors of the Hanford or Chernobyl type, neutrons are slowed by collisions with the carbon atoms in graphite. Graphite is composed of tightly bound carbon atoms in honeycomb-like planes. The high velocity neutrons knock the carbon atoms out of their stable places in the lattice. The displaced atoms then perch in high-energy positions between the planes.

If the reactor operates at a sufficiently high temperature, the carbon atoms find new positions in vacancies in the lattice or at the ends of the planes. At a lower operating temperature, energy slowly but steadily accumulates in the graphite in the form of displaced carbon atoms. Then a rise in temperature can set off a release of stored energy in the graphite, which, of course, further intensifies the original overheating problem.

Another important question is whether fission increases or decreases when the temperature rises. A large number of factors interact here, but if the sum of their effects causes increased reactivity when the temperature rises, there is danger. If an increased temperature, at every power level, tends to shut down the reactor, the design is safe in respect to the direct effects of temperature.

Reactors have to be cooled, and the circulating coolant is often water. Water in graphite reactors has two opposite effects on the fission process. First, because neutrons are slowed by colliding with water molecules, the presence of water raises the efficiency of the fission process. The loss of water thus decreases fission. But in addition, water absorbs slow neutrons much more effectively than the graphite does. The loss of water, therefore, increases the number of available slow neutrons and thus speeds up the chain reaction. Which effect is dominant depends on the amount of graphite. In reactors with a large amount of graphite, like those at Chernobyl and the wartime Hanford reactors, the water has a more important effect in absorbing slow neutrons. Thus, loss of water causes a self-intensifying increase in power and temperature.

In the Soviet graphite reactors, the water is circulated under pressure at a temperature close to the boiling point. That produces a maximum amount of steam to run the turbines, which generate electricity. If the boiling water is converted to steam within the cooling system, or if the water is lost in an accident, a void is created. The void increases the number of slow neutrons available and can result in dangerous prompt criticality.

One additional problem was recognized from the beginning days of reactors. When a reactor is shut down, the fuel continues to release heat (initially about 7 percent of the energy that was being produced at the time of shutdown) for a considerable period. That afterheat must be carried away, or it can cause a meltdown of the nuclear fuel. Therefore, circulation of a coolant needs to be ensured even after a reactor is shut down.

The five potential causes of trouble—xenon poisoning, Wigneritis, reactivity increased by a rise in temperature, reactivity increased by loss of coolant, and afterheat—were thoroughly investigated and understood by 1954, when the First Atoms for Peace Conference publicized this information before an international forum.[2]

Long before that conference, the Advisory Committee on Nuclear Reactor Safeguards had acquired a nickname: "The Advisory Committee on Reactor Prevention." We had recommended that the general public be excluded from an area whose radius would be determined by the operating power of the reactor. (The restriction was later removed and replaced by the requirement of a containment building sufficient to prevent the spread of radioactivity in all foreseeable conditions.) Our recommendation clearly applied to future reactors, but the question arose as to whether it needed to apply to the Hanford reactors built during the war. After all, those reactors had operated without serious difficulties for four years. If the exclusion rule applied to them, the public could not use the rich agricultural land within 5 miles of the plant.

Our committee was asked to consider the special question of the Hanford reactors. After careful discussion, we concluded that the restriction should indeed be applied. The AEC did not like our recommendation but did not feel it could be disregarded completely. So the AEC appointed a second committee, mainly made up of engineers and administrators. The chairman of the new committee, Rogers McCullough, asked that it be allowed to sit in on our discussions. That, of course, was agreeable.

After a year had passed in study, the second committee came to the same conclusion: The wartime Hanford reactors could not be considered completely safe.[3] Since the two committees agreed, there was no point in duplication. We made a joint recommendation that the groups be united, and a few years later, about the time that secrecy restrictions were lifted, the Advisory Committee on Reactor Safety was established by law. More than thirty years later, the information gathered by the early committees has become pertinent in a tragic manner.

The four reactors at Chernobyl, each capable of a 3,200-MW thermal output, are a second generation of the first Soviet design and have a massive amount of graphite. As a result, various portions of the reactor may have differing temperatures, neutron fluxes, and xenon concentrations. That makes the reactor difficult to operate. The greatest danger is that an increase in voids in the cooling system can lead to prompt criticality.

At 1:00 A.M. on April 25, 1986, the operators of Reactor No. 4 began to prepare for an experiment related to sustaining emergency electric power. The people conducting the experiment were electrical experts, unfamiliar with the complexity of the reactor and working under strict time limits. The intent was to conduct the experiment at about 20–30 percent power. By 2:00 P.M. they had reduced the power and had disconnected several of the safety devices, including the emergency core cooling system (which was never turned back on). However, the plant dispatcher requested a delay of the shutdown, so the reactor continued in operation until 11:00 P.M. The operators, attempting to overcome the xenon poisoning, shut off the automatic power regulator. Two hours later, at 1:00 A.M. on April 26, they managed manually to stabilize the reactor at about 7 percent of the normal output. They started the experiment.

At 1:03 and 1:07 A.M. two additional cooling system pumps were brought into use. The cool water decreased steam production in the turbines (which power the circulation of water), and at 1:19 A.M. the water level in the pumping system began to drop. The operator increased the water flow rate fourfold to correct the problem. As still more cool water reached the core, the pressure in the system dropped further, and the adjustment rods automatically began to rise to increase the heat by increasing fission. The operators assisted the process manually by moving the rods up further.

At 1:22 the operators sharply reduced the water flow and, in accordance with the experiment, at 1:23:04 A.M., closed the valves on one of the turbines. The flow rate fell further, and voids began to occur in the cooling system. The Chernobyl reactor is more sensitive to voids when operating at low power with cold dense water and without any voids. In this condition, even a rise in water temperature increases the reactivity. With cold water filling the reactor, with no voids present, and with most control rods withdrawn, the water in the Chernobyl reactor was worth three dollars. About thirty seconds later, the power surged. The xenon poisoning had been overcome, although probably not uniformly throughout the reactor. Increased reactivity caused localized overheating. The self-intensifying process of increasing voids, increasing reactivity, increasing temperature, and again increasing voids had begun.

At 1:23:40 A.M. the operator ordered insertion of the emergency control rods. He was too late. The reactor power increased to 50 percent of full operation and then began to rise rapidly. Calculations suggest that at least thirty times full power was attained—probably much more for a short time. Reactor No. 4 was a runaway. Less than twenty seconds later, at 1:24 A.M., extremely high steam pressure knocked askew the 1,000–ton reactor cover, the principal means of containment, and destroyed the lower third of the reactor. Water from the broken pipes combined with the superheated zirconium covering the fuel and released hydrogen. A few seconds after the steam release, a second explosion occurred, probably from the hydrogen combining with oxygen. Large burning chunks of graphite were blown out of the building and set fires, some in and around the Reactor No. 3 building.

Three of the five potential dangers identified before 1954 by the first Advisory Committee on Nuclear Reactor Safeguards were present in the Chernobyl reactor: xenon poisoning, increased reactivity from voids in the cooling system, and afterheat, which, because the entire reactor was in ruins, could not be controlled. During the next week, helicopters dumped a total of 5,000 tons of material—boron carbide (to absorb neutrons), lead and dolomite (to absorb heat and keep oxygen away from the burning graphite), and sand and clay (to bury the smoldering radioactive remains). Radioactive emissions decreased for six days after the accident, but then the radioactivity increased sharply, perhaps because the heat was transferred from regions of higher temperature or conceivably because of some chemical reaction. After four more days, the activity tapered off rapidly. No further emissions are likely.

The Soviets are modifying their graphite reactors that are similar to those at Chernobyl. Fixed, unremovable absorbers are being installed, and the fuel is being enriched with uranium-235 in order to reduce the effect of voids on reactivity. The hope is that loss of water will no longer add more than one dollar to the reactivity. In the West, there are no power reactors in which a void in the cooling system can produce such prompt criticality. In 1950 the United States decided that reactors of that kind were not safe enough.

In the thirty-year history of electricity-generating reactors, two other major accidents have occurred. The first, in 1957, involved an early, uncontained graphite reactor at Windscale in Great Britain. Windscale operated at a temperature too low to prevent Wigneritis, but the potential danger was recognized. The accident occurred during an attempt to correct the problem. The operators brought up the temperature, but the accumulated energy was released with unexpected speed. The graphite ignited, and some time passed before the fire could be extinguished. Radioactivity was released and deposited as fallout on nearby grazing land.

A variety of the radioactive materials in a reactor can be vaporized and emitted. Three isotopes—iodine-131, cesium-137, and strontium-90—are of particular concern, because they are incorporated in living things.[4] Radioactive noble gases have an almost negligible effect, because, being chemically inert, they cannot be assimilated into living tissues. The Soviets report that 20 percent of the radioactive iodine, 10 to 13 percent of the cesium, and 4 percent of the strontium in the Chernobyl reactor actually escaped. The biologically damaging fallout from the Windscale accident seems to have been about one-thousandth that of Chernobyl. The British impounded the milk containing iodine-131, converted it into dried milk, and stored it until the radioactivity had decayed away. No adverse health effects were observed.

The second accident occurred near Harrisburg, Pennsylvania, at the Three Mile Island (TMI) plant in March 1979. Both the reactor design and the containment system were different from the Windscale and Chernobyl reactors. At 4:00 A.M. the reactor automatically shut down because a pump in the cooling system failed. Unfortunately, a valve stuck in the open position caused the reactor to lose water. The afterheat raised the temperature, and the automatic cooling system injected more water. Converted to steam, the water produced thumping noises. The operators misunderstood the problem, manually shut down the safety system, and turned off the water. The afterheat then melted the surface of fuel rods, setting off a zirconium reaction, which freed hydrogen. A small explosion occurred when the hydrogen mixed with air. However, the containment building—several feet of steel-reinforced concrete—was not damaged. The amount of radioactive iodine and cesium that escaped was one-millionth that released at Chernobyl. If the highest estimates of risk for radiation are compared with the risks on highways, the people near TMI were much safer staying home.

In sharp contrast to Windscale and TMI, the accident at Chernobyl was truly dangerous. The effects of a worst case accident have been variously predicted; some speculations mention 100,000 to 300,000 deaths. It is inexcusable that the media circulated those unfounded assumptions during the early days following Chernobyl.

The accident at Chernobyl unfortunately did involve loss of life. In the first days after the accident, 203 people, all of them exposed at the reactor site, were hospitalized with radiation sickness. At last report, thirty-one people—plant employees and firefighters who remained on site trying to limit the damage—have died. In August when the report was issued, thirty patients were still hospitalized but were expected to recover.

Soviet civil defense preparations may have been employed to moderate the effects of the accident. Cloud-seeding with silver iodide was used to prevent rain in the critical areas. Immediately after the accident, the people of Pripyat, the city serving the Chernobyl complex, were told to minimize their time outdoors and to keep their windows closed. Prophylactic iodine treatment was given at nursery schools, kindergartens, and schools. On April 27, when the radiation level began to rise, the 45,000 Pripyat residents were evacuated in three hours. Later, about 90,000 people within 20 miles of Chernobyl were evacuated.

Pripyat residents received between 1.5 and 5 rem of radiation, but exposure among a few other evacuees may have reached 40 rem, which is still less than half the dose that produces radiation sickness.[5] The important question is what effects the accident will have on life expectancy. Perhaps as many as 1,000 premature deaths among the many millions exposed in the Soviet Union may occur during the next seventy years. (In the same population, a far greater number will die prematurely from smoking cigarettes.) There is no indication that anyone outside the Soviet Union will be adversely affected.

The commodities market, which rose briefly and then resumed its usual level, correctly interpreted the effect of Chernobyl on the harvest. The Soviet report announced that grain, after curing in warehouses, will be used for food, animal feed, and planting. Some milk was quarantined during the time it showed an excess of iodine-131, as were leafy green vegetables, spices, and tea grown near the plant, and some meat has been put in cold storage for later distribution. Meat will continue to be checked for radioactivity for some time.

The fallout is expected to have severe effects on the water plants and fish in the cooling pond at Chernobyl. Fallout in the Pripyat River may have some effects on marine ecology. No further concentration of radioactivity in the water system is expected.

Chernobyl had one salutory consequence. It produced a departure, at least to some extent, from the old Russian practice of secrecy. Neither czars nor commissars have ever publicly admitted to a disaster. Gorbachev acted in a much more open manner; the results of a thorough investigation of the accident were disclosed. That represents real and important progress.

In the wake of the Windscale accident, the design of reactors was altered to avoid the possibility of Wigneritis. The TMI accident, while harmless, cost more than a billion dollars. Afterwards, the nuclear industry established its own safety analysis group, which regularly evaluates each plant and provides safety training for operators. By 1986, the rate of occurrence of malfunctions had dropped by more than half.

Chernobyl has reemphasized the problem of safeguard systems that can be bypassed or shut down by operators. If, at TMI or at Chernobyl, the operators had been unable to shut down the safety systems until three or four truly knowledgeable reactor authorities were present—even if it took a week to assemble them—the world would never have heard of those places.

Chernobyl, like many reactors throughout the world, was built rather close to a water table. Had a meltdown penetrated a relatively small number of feet downward, the water could have been contaminated. Considering the potentially serious consequences, attention should be given to locating reactors where such contamination would be avoided even in the worst circumstances.

It also seems worthwhile to reopen an old question. The importance of containment structures has been demonstrated. However, containment buildings such as those used in the free world are expensive. It would seem reasonable to look more carefully at the cost of locating reactors underground. Studies clearly indicate that siting a reactor approximately 100 feet under the earth's surface in loose soil would minimize dangers from any explosion.

Designs for an improved afterheat exchange system are being considered. The system would transfer the afterheat directly to the surrounding medium. That would do away with the need for an external water supply or external energy sources and would make it impossible to interfere with the heat transfer. Safety can be provided more reliably by improved technology than by regulations.

The accident at Chernobyl occurred in the kind of reactor that was considered unsafe almost four decades ago. The consequences of a truly terrible accident appear to be much more limited than the effects that were imagined and publicized. But we must search for reactors with maximum inherent safety and locate them in such a way that, even in the worst imaginable case, neither air nor water will be contaminated.

Nuclear energy is needed to ensure economic well-being throughout the world. The possibility of having clean, secure, safe nuclear energy exists today. But that can be accomplished only if the lessons of all nuclear accidents, including those of Chernobyl, are learned.

General Secretary Gorbachev has called for greater international cooperation on the safety of nuclear reactors. The requirements for cost-effectiveness seem to have been emphasized more thoroughly in the Soviet Union than in the free world. Safety has had priority in the West. We should work together on both of those important requirements. Active cooperation could become the means for increased mutual understanding.

21

The China Syndrome:
A Cautionary Tale

The expression "China syndrome" had its origin in an inappropriate joke of the late 1940s: What would happen if a reactor would melt down and its containment vessel were breached? The answer: The fuel would end up in China.

In the United States during the 1960s a reaction against technology became widespread. It bears some resemblance to a development in fifteenth century China. That event was real and can appropriately be called the "China syndrome."

In ancient times, China was far ahead of the rest of the world in technology. Ch'in Shih Huang Ti, the man who created the Chinese empire about 200 B.C., connected the defensive structures along the northern and western borders of China into a single great wall. That wall now stands as a monument to peaceful defense. It actually served to exclude the savage Mongol tribes from China for 1,400 years. While invasion destroyed the Roman empire, China, the Middle Kingdom, remained peaceful in its imagined position halfway between heaven and earth.

Even before Ch'in Shih Huang Ti, the Chinese had invented silk. That invention remained a secret process for millennia and contributed to a trade imbalance in the West. Tiberius, the emperor of Rome in the first century A.D., forbade men (but, being a wise politician, not women) to wear silk so as to decrease the "specie hemorrhage to the East."[1]

The Chinese invented printing, gunpowder, rockets. They also invented a harness that allowed horses to pull loads. The harness used in the West was appropriate for cavalry but choked a horse when used for draft purposes. Many centuries later, around A.D. 1000, when the Chinese-style harness was introduced in the West, the horse became most important in European agriculture for plowing. A beneficent revolution followed as a result of an imported improvement in technology.

The list of Chinese technical accomplishments is long. In A.D. 600 a canal was built connecting North and South China. It strengthened the unity of that huge country through improved transportation. The Grand Canal did a great deal to prevent local famine. Nothing comparable to that canal was built anywhere in the world until more than a thousand years later.

One of the most remarkable of ancient Chinese devices was the compass, which was used before the birth of Christ. By the middle of the thirteenth century, it had enabled China to establish trade with countries ranging from the East Indies to India to the Persian Gulf to Madagascar and the east coast of Africa. Modern archeologists exploring the ancient remains in east Africa have found that from the tenth century onward, Chinese porcelain is present in great quantity.

During the first decade of the fifteenth century, the Chinese amassed a powerful fleet—3,000 ships, including 250 galleons roughly five times the size of Portuguese vessels that sailed later in the century. The Chinese ships, with crews of more than five hundred men, also were armed with cannon more sophisticated than the contemporary arms in Europe. In the years following 1405, Cheng Ho, the eunuch admiral of sixty-two huge ships and tens of thousands of men, embarked on seven expeditions. On the last trip, begun in 1431, he visited twenty realms and sultanates, including many along the east coast of Africa.

The Chinese came as traders, exchanging gifts of gold and other treasures for exotic curiosities. One serious military action during those journeys was recorded. The King of Ceylon demanded excessive presents and eventually sent his troops to burn and sink Cheng Ho's ships. Cheng Ho captured the king and his court, fought his way back to the coast, routing the army on the way, and took the prisoners back to China. There they were received kindly and sent home again after agreeing to choose a relative of the king as his successor. Armed might was used discreetly by the Chinese.

The motives for those trips, for the most part, can only be guessed. Apparently, one of the purposes was to impress the world with the idea that China was the leading political and cultural power. There was also a proto-scientific purpose. Cheng Ho was accompanied by numerous physicians, astronomers, geomancers, and naturalists. A Chinese writer of the period says:

> In my youth, by reading books such as the *Record of Strange Countries*,
> I learned the vastness of the surface of the earth, the differences in customs, the diversity of human beings, and the variety of natural products—
> which are all truly astounding, lovable, admirable and impressive.

During the same century, when Henry the Navigator established his famous laboratory in Sagres, Portuguese ships started to explore the west coast of Africa. But long before that the Chinese had begun to explore the globe. The distance from Portugal to west Africa is small; the distance from China to east Africa is almost halfway around the world. In the middle of the fifteenth century, there was every reason to guess that China would rule the world.

But that splendid beginning came to naught. This is how history records and interprets the change. The Chinese wall was not a hundred percent effective. Around A. D. 1200 China was conquered by Genghis Khan. His descendants ruled China for more than a century. In the end, the Chinese revolted, and a Chinese adventurer-turned-emperor founded the Ming dynasty.

In the Ming court, and also at earlier times, were two opposing groups of considerable power. One group contained the Confucian bureaucrats who had earned their position by literary examinations open to everyone. The high officials, the Mandarins, appointed on the basis of their knowledge of the great old philosophers, were the custodians of ancient Chinese values and literature. The other group, the eunuchs, purchased as children by the emperor, were completely loyal to his person. They were interested in technology and on the whole were a much more innovative lot.

The Confucian scholars and the eunuchs represented two cultures that may have differed more widely than the two cultures of literati and engineers in the present-day West described by C. P. Snow. The emperor used the less refined eunuchs to check on the august and highly polished Mandarin governors. The eunuchs occasionally discovered an irregularity. Not much love was lost between the two groups.

When Genghis Khan moved the capital of China to Peking, supplying food to the new capital became a problem, because the Grand Canal did not reach that far north. For a number of years, while the canal was being extended, food was brought in by sea. The eunuchs led the political party that supported the development of the navy. The party of the bureaucrats, with their connections to the rural landlords, bitterly opposed that development.

The Ming emperors varied in their attitudes to the navy. In 1424, an emperor came to the throne who ended the expeditions, but he died within the year, and the next emperor sent Cheng Ho off on his last mission. That emperor died in 1435, however, and the policy was again reversed, this time permanently. An edict in 1448 made it clear that the fight between the two groups had been won by the Confucian scholars. Not much is known of the struggle, or even of the details of Cheng Ho's journeys. Once the bureaucrats consolidated their power, they burned the books describing the overseas adventures.

Is that the whole story? I doubt it. Perhaps there are deeper factors that govern the fate of individuals or empires. I have asked many of my Chinese friends. None of them would even guess. The events and people that brought about the change covered their own tracks.

The result of the edict, however, is clear: All exploration stopped. Eventually the ships were destroyed. The Ming emperor said in effect: In the whole world, the only thing that matters is the Middle Kingdom. All the real needs of food and clothing, including goods of fine craftsmanship, are available at home. Why spend money on foreign objects? The funds spent on the navy could be better used for water-conservation projects or agrarian financing for ever-normal granaries.

Most remarkably, from that time on there were no more great technological developments in China. China had had enough of innovation, and the leader of the world at that time started to turn into an underdeveloped country. During the same period, the ancient Chinese ceramic art blossomed, and beautiful multicolored porcelain of outstanding excellence was produced. A choice had been made for culture and against technology.

Four hundred years later, East and West met. The Chinese emperor accepted with disdain the barbarian Englishmen bearing tribute. However, it soon became clear who possessed the real power. A series of tragedies for China followed: Western intrusion, the Opium Wars, and internal disorders like the Tai Ping rebellion, which alone cost millions of lives. For two centuries the Middle Kingdom was a place closer to hell than to heaven.

The West began technological development just at the time when the Chinese stopped being interested. Westerners in various ways—

some horrible, some helpful—accomplished the exploration and colonization of the world. The peculiar situation in China, the unexplained recoil from technology, is one of the most significant events in recorded human history. This was the real China syndrome. China invented it. The United States is imitating it today.

Not many years ago, Americans landed on the moon. The NASA budget was cut soon afterward. The belief seemed to be that public interest should be confined to what happens in the United States. Technology, even science, came to be seen as irrelevant, perhaps even dehumanizing. Today, four-fifths of American high school students take no science or mathematics courses beyond tenth grade. There can be no doubt that currently a larger proportion of high school and university students in the Soviet Union and Japan concentrate on science and technology.

Throughout the history of the United States, increasing productivity has been the pattern. In 1980, productivity declined for the first time. There are many reasons for that. The energy crisis is only one of them. Young people during the 1960s turned away from engineering and science. Their productive years have been directed toward other activities.

In China, four hundred years passed from the time of the original imperial edict to the time when its harsh consequences became evident. In the United States, some effects have become noticeable in less than twenty years. The modern world changes at a rapid rate.

There are relatively few poor people in the United States, but more than three-quarters of the people on earth are wretchedly poor. The American standard of living and American security are due to advances in technology. At present, a historic change is occurring in the world: The globe is being engulfed by the industrial and scientific revolution. The poor nations are not confused about the source of well-being for their peoples.

As technology spreads and increases, there will be new common interests, new cooperation, and new conflicts. Modern development, modern technology, has the inescapable consequence of increasing interrelationships over the globe. Interdependence will occur within the lifetime of many of those living today.

When the actions in one country influence what happens in another country, then complete national independence is gone. Those changes cannot occur without some kind of world order, although not necessarily world government. One practical time-honored way in which to deal with such situations is to impose imperial control by force. The fate of nations in Eastern Europe following World War II and more recent examples, such as Afghanistan, indicate that such control has not gone out of style. The Soviet Union is prepared to enforce its domination using military technology.

The postwar recovery of Japan and Germany, with American help, shows that there is a different possibility. If technical development in the United States were resumed, then the order required in an interdependent world might be produced not by force but by reason. Democratic arrangements for worldwide collaboration might then occur, not on command but by agreement. This is something new and miraculous. It will not happen if the United States loses leadership in technology, if the nation follows the China syndrome. That syndrome leads into the abyss.

How long will a nation dependent on science and technology flourish if the majority of its people cannot make technically informed decisions? If most people entering the job market are scientific illiterates? We are living off our technological capital. Clearly, it is about to run out. An out-of-control reactor producing a melt-down headed for China is a fantasy. A nation moving in the direction of China's past error is not.

Chinese history is vitally important to our future. Technology alone does not suffice. The eunuchs alone could not have governed China well. Neither could the Confucian literati. The two cultures are complementary. They must not remain separate. A synthesis is necessary. It will not be easy. It is, however, a requirement for the survival of our civilization whose roots reach down to the Renaissance, to Sagres, and beyond.

22

Unmanned Vehicle Systems

Few men can fight; no machine can think. But a machine can fight, gather information, and carry out orders, and a man at a remote location can do the thinking. A light beam or similar radiation can carry the necessary messages between them with practically no delay. Such a combination of man and machine is called a remotely piloted vehicle, or RPV.

American technology could increase our military power by introducing unmanned vehicles. A primitive form of such a vehicle is the cruise missile. But the cruise missile is guided by a preconceived program. A missile or vehicle with the ability to collect and send information to a command post, which then can return the appropriate instructions to the vehicle, is much more effective. RPVs could be particularly important in defense. The exchange of information and instructions is especially easy in purely defensive operations conducted at short distances.

Unmanned vehicles, in contrast to conventional defensive military equipment, are small and cheap. They break the deplorable trend toward bigger machines of greater complexity and higher cost that leads to scarcity of equipment. In addition, unmanned vehicles are expendable, not only because they are cheaper, but because no human life is directly involved. RPVs, however, do require computers able to handle a great deal of data. There are few fields of military technology in which the free world has a real advantage. Computers and microelectronics are the most important.

RPVs become more effective as they become more flexible and responsive to human commands. An improved version of every human sensory ability can be included in such vehicles. Equipment on RPVs can measure acceleration, register sound, send back a visual image, get an orientation reading—collect every kind of information a man can perceive and more. Too strong a signal shocks, blinds, or deafens a man. Machines easily can be made more resistant, more adaptable, and at the same time more sensitive.

The sensory addition that I particularly like is laser eyes. Human vision operates poorly at night, and the light reflected by a thin cloud interferes with our perception of the light reflected by an object within that cloud. Laser eyes not only see well in the dark but, with an appropriate timing system, can see through thin clouds. The RPV can send out an extremely short and intensive laser beam—lasting less than a billionth of a second; the actual train of waves will have a length of only a few inches. When the waves hitting an object in the cloud return to the laser eyes, a sudden increase in the signal will occur, which can be easily distinguished from the scattering produced by the droplets.

All of the information collected can be sent back at the velocity of light to a control center where a crew can evaluate it. The communication link may be produced by lasers or by microwaves which are better in penetrating thick clouds. RPVs can be connected directly with a control center, or they can be linked through a space satellite. In the latter case, they can be operated at distances beyond the horizon.

Unmanned vehicles can operate in the air, on land, and under or on water. Underwater use is more difficult, because electromagnetic waves do not propagate there. Blue-green lasers of high intensity, however, penetrate to considerable depths; at greater depths sound waves, called a sonar link, can be used. RPVs can be used for reconnaisance, for attack, or for defense. Unmanned tanks would be both small and efficient. The Israelis used remotely piloted aircraft against the Syrians in 1982. The RPVs won by a one-hundred-to-one margin.

Clearly, unmanned vehicles will play an increasingly important role in military affairs. The United States badly needs to strengthen its conventional forces, but sufficient funding to do so may not be available. Therefore, it is particularly important to use our brains and to use the technology in which we have an advantage. Those considerations point to RPVs, not just the programmed cruise missiles but the flexible vehicles of the varieties mentioned.

What stands in the way of implementing such a program? I suggest the problem lies in the fact that of all materials known to man, the one with the greatest inertia is the human brain. The one exception is that a collection of human brains such as is found in the faculty of a university or in the Pentagon shows even more inertia. But what the Pentagon will not do, private industry can undertake.

Many jobs in our society are dangerous. For example, crop-dusting requires pilots to fly at very low altitudes. Unmanned vehicles could do the job without risking human life. A crash might cause a fire, but if the RPVs were equipped for firefighting and worked in pairs, no harm would be done.

RPVs could do many tasks more efficiently. In the Western United States, dry summers often produce severe forest fires. The early detection of forest fires is important, and unmanned vehicles could provide an excellent patrol system. In addition, firefighting equipment could be operated from unmanned vehicles, spraying otherwise inaccessible areas with fire retardants or water. RPVs might even be equipped to create firebreaks. Quite possibly, RPVs could save the lives of many firefighters, stop forest fires earlier, and save many acres of woods.

Not only could RPVs protect the forest, but they could be very useful in putting out residential fires. A few years ago a large area in Los Angeles burned. The disaster could have been avoided by spraying the adjacent roofs with fire-resistant materials. That was a huge job, however, and had to be done quickly. The fire department could not do it with the available means. The fire burned out of control and destroyed many beautiful homes.

Unmanned vehicles could also be of assistance to police departments. Safety in big cities might be improved by the use of RPVs equipped to gather information about after-dark activities and transmit it to a receiver in a central manned station. The laser beams could be of infrared or ultraviolet light, which is not visible to the unaided human eye. Using that technique might make Central Park in New York safe and enjoyable again.

Some tasks are feasible only for an RPV. For example, gathering manganese nodules on the ocean floor can be accomplished only by remote control. Drilling for oil on the continental rise a mile under the surface of the water also could be done by unmanned vehicles.

If manufacturers and their customers could demonstrate the effectiveness of these devices in our daily life, perhaps then even the Pentagon would take notice. There is a certain capitalistic bias in that statement. If I were to apply my law of inertia to a collection of human brains in industry, I would have to conclude that my proposal is as difficult as trying to sell the program to the Pentagon directly.

However, I hope that private enterprise is just a little more clever than other types of organizations, because in industry unusual stupidity is punished promptly by one's competitors. Indeed, private enterprise might do a lot of things that the government does not manage to do or does not manage to do well enough.

The 1980s have begun with a dangerous military lead held by the Soviet Union. Unmanned vehicles could help to remedy that situation.

23

The Lunar Laboratory

Thinking about a lunar colony has occupied many fine science fiction writers, including H. G. Wells, for a considerable length of time. I too have been interested in the idea, I hope on a realistic basis. Today, a lunar laboratory seems to make sense on scientific, technical, and even economic grounds. I have not proceeded much beyond general estimates. The details are offered partly for illustration and partly to clarify a few differences from generally discussed ideas.

What form should a lunar laboratory take, and what projects should be attempted? In the initial stage, only a minilaboratory could be considered. Such a laboratory would be staffed by about a dozen people. A few months after Sputnik, I was asked an interesting question about space during a Senate hearing: Should there be any female astronauts? I answered that all astronauts should be women—they weigh less and have more sense. Intelligence seems to be better packaged in women. But nowadays, with affirmative action measures, I have to modify my recommendation. There should be six women and six men to staff the first minilaboratory.

The workers would be rotated back to earth after a limited number of months on the moon. Spending an extended length of time in a region of low gravity leads to decalcification of the bones. On the surface of the moon, gravitational acceleration is one-sixth that of the earth. The lunar laboratory people would have to work inside space suits. Whether a person's bones carry less mass and more gravitational acceleration, or more mass (in the form of a space suit) and less gravitational acceleration may not make much difference in terms of stresses on their skeletons. Probably people can work on the moon for longer than three months without incurring physiological problems. The duration of their stay might be a year. The real limitation may well be psychological.

Locating the living quarters of the lunar colony in a crater has advantages. Because of the periodic intense radiation produced by flaring sunspots, laboratory workers will need considerable readily available shielding. If the laboratory is built over the edge of a crater, the workers would be able to duck into a shadow quickly. More than one crater may be needed, because any single location will be exposed to heat or cold for too long a time.

It would be even more ideal to place the living quarters inside a cave in the wall of a crater. The temperature in lunar caves, while cold, is more moderate, because an average of the lunar day and night temperatures prevails inside a cave. Living quarters in a cave would save energy and provide extra insurance against unusual levels of solar radiation.

Further advantages are gained by placing the base near one of the lunar poles. Both sun (and heat) and shadow (and extreme cold) would be within easy access. Similarly, from such a point the earth can be seen, which is useful for communications. Yet nearby areas are shadowed from the earth which is preferable for astronomical observations.

There seem to be three appropriate craters close to the south pole. Craters are scarcer near the north pole. Starting from the south pole during the safe period of sunspot minima, when no solar flares are expected, the workers could spread out and establish various projects.

A great deal of material will be required if a dozen workers are to work effectively on the moon. My estimate is that a minimum of 20 tons per person will be required each year. Only a small fraction of that weight will have to do with supplying the worker's physical needs for food and water. Obviously, their water would be recycled, a technology already well developed. The main fraction of the weight will go for energy (crucial to their survival as well as their work) and equipment for building and research.

The suggestion has been made that we learn to grow food on the moon so that the colony could be self-sustaining. That attempt should be postponed. The valuable opportunity of being on the moon seems wasted in mastering agricultural activities. Man's development on earth has been from hunter to farmer to technologist. On the moon, the first act should be technology. Agriculture can wait until the United Lunar Colonies formulate their Declaration of Independence. Hunters on the moon will be out of luck.

The main expense of the lunar laboratory is apt to be the transportation of material to the moon. That expense is what causes me to set the size of the lunar colony at twelve people. Flying the shuttle costs much more than was originally expected: Each trip carrying 30 tons of payload costs about $200 million. Considering the additional rocket fuel required to carry such a payload all the way to the moon, the cost may easily triple.

At a cost of $600 million for 30 tons, the annual expense of transporting 240 tons to the lunar laboratory (twelve people times 20 tons apiece) would be $5 billion. That figure does not include the cost of preparatory research and fabrication of the materials to be delivered. The development of the necessary technology (including transfer vehicles) will take about three years and several billion dollars. But combining all the expenses (including transportation of the lunar laboratory workers), a total annual budget of $10 billion might suffice. Remarkably enough, that is not a great sum in comparison with past NASA expenditures.

Great scientific and industrial benefits can be expected from a lunar laboratory, but the first priority should be to create a station to refuel rockets with oxygen and possibly hydrogen. Rockets can be accelerated to a velocity of approximately 4 kilometers per second (km/sec) by an amount of fuel comparable to the weight of the payload. Each additional 4 km/sec acceleration requires a doubling of the fuel. Thus, leaving the earth (which requires 11 km/sec) plus landing on the moon requires a great amount of fuel; further maneuvers that require fuel become very expensive. Takeoff from the moon (which requires 2.4 km/sec) and orbital velocity around the moon (1.5 km/sec) are comparatively cheap if lunar-produced fuel becomes available.

Even if a few years were required for its establishment, the benefits of a refueling station outweigh all other advantages. It would make the lunar laboratory much more practical by making commuting back and forth cheaper. Lunar refueling will postpone or eliminate the need for nuclear-fueled rockets. The refueling station is the real basis on which the future of our space enterprise depends.

The specific importance of the moon is that it contains plenty of "green cheese" that can be turned into useful products. Fuel for space travel could be obtained from either of two sources. Lunar rocks are essentially oxides. Those oxides should be selected from which oxygen is most easily liberated. Mechanical energy will be needed to crush the rocks so that the oxygen can escape more easily.

An alternative to using lunar rock is to use lunar dust (regolith) since it already has the proper physical form. Regolith consists of particles one millimeter to one micron in size and covers most of the lunar surface. However, its chemical composition makes it a less desirable source of oxygen, and mechanical means will be needed to collect sufficient quantities. On the other hand, regolith is a good source of hydrogen, present at a concentration of 10^{19} atoms per cubic centimeter (less than one atom in a thousand) in a lightly bound form. The hydrogen has been deposited by the solar wind. Thus, fuel for a hydrogen-oxygen rocket could be made available from lunar material. Even if only oxygen is made available, large savings will be made. Transporting hydrogen from the earth involves considerably less weight than transporting oxygen.

Liquefaction of the fuel will not be a problem, because the lunar nighttime temperature is sufficient to accomplish the process. In the case of hydrogen, a little more effort is needed for liquefaction. However, obtaining the fuel will require great amounts of energy. In addition, the lunar colony will need some energy to conduct its work as well as to survive. Massive amounts of energy for fuel production and a far smaller quantity for general use are needed. Two sources seem obvious—solar energy and nuclear energy.

One method of heating the lunar rock would be to focus solar light on a small area. Temperatures up to 3,000° Kelvin—which are certainly sufficient—can be achieved. The difficulty is that the system of mirrors required to reach those temperatures is apt to be heavier than a reactor and therefore more expensive. The question is not whether fuel can be produced using solar energy but how much it would cost.

One advantage of solar energy, particularly solar electricity, is that it can be made available in widely distributed locations on the moon. However, if solar cells are used, that energy supply will disappear during the fourteen days of solar night, which occur to a lesser extent and in a different way near the poles. Maintaining a continuous supply of energy will require batteries, which being heavy are, again, expensive to transport. Ultimately, ways may be found to make batteries out of lunar materials.

One suggestion that arose a few years ago in connection with solar energy was that it should be collected on a satellite, converted into microwaves, and beamed back to earth for reconversion into electricity. The idea seems unlikely to become economically feasible. But if practicality can be approached, the moon would be a better location for the initial conversion of solar energy.

Using a thermoelectric source based on the temperature differences on the surface and a few feet below the surface should also be considered. Although the sign of the difference changes, that temperature difference is available day and night.

A nuclear reactor would provide a good source of large amounts of heat. Transporting the heavy shield for a reactor would be unnecessary. Only a specially constructed core of a nuclear reactor would have to be sent. The shielding material could be made of lunar rock. In fact, the whole reactor could be built into a lunar cavity in such a way that it would be well shielded. Lightweight excavation equipment will be essential for that task, as well as for several other purposes.

Cooling and maintaining a lunar nuclear reactor would present problems. Therefore, if a nuclear reactor is used at all, it should be specially constructed in a simple manner, mainly for use in obtaining the fuel for rockets. The reactor must be safe and sturdy; in case of malfunction, it should be replaced rather than repaired. Those questions, of course, will require research.

A further requirement is to keep the oxygen that is produced free of radioactive contamination. Oxygen itself will not give rise to disturbing long-lived radioactivity, but materials associated with the oxygen do. Methods would have to be developed to eliminate traces of them. While in principle this should be possible, carrying it out with remote-control apparatus would not be easy. Experiments to develop appropriate processes would have to be carried out beforehand on earth using lunar materials.

Ultimately, however, both solar and nuclear sources of energy should be available. The nuclear source could provide massive amounts of heat; the solar source would offer a modest amount of electricity in the areas where special projects might be located.

As a refueling center, the moon could serve as a jumping-off station for the exploration of the whole planetary system. One of the great advantages of a lunar colony is that it would make the whole space program considerably less expensive. Having to bring all fuel from earth and having to carry the fuel to overcome solar gravity is extremely expensive. Refueling on the moon would lead to dramatic savings. In the twenty-first century, laboratories on small man-made moons, on moons of other planets, and eventually on planets themselves might be established.

One cannot say that the refueling function of the lunar colony would pay for itself, because exploring the solar system may give us nothing except knowledge. That knowledge may provide us with enormous advantages, but knowledge—unless we regard it as intellectual energy (corresponding to infinitesimal fractions of micrograms) and invoke the $E = mc^2$ equation—has no weight.

After establishing the refueling station, the next task I would propose is not to go on to the solar system but to explore the moon itself. We know almost nothing about the geology of the moon, a branch of science known as selenology. A little is known about what is a few feet under the surface, and the rest is inference. Obtaining cores down to a few thousand feet would help explore the history of the moon. Taking corings at great depths may turn out to be somewhat expensive and will hardly be feasible during this century.

Knowledge about crater formation would also be gained. Most craters were probably made by meteoric impact, but a few may have been made by volcanic eruption. Furthermore, those two phenomena are probably not entirely independent. A large meteor impact may well have effects on the lava layers that, in some time sequence, show up as volcanic action. The relationship between the maria, the flat "seas" of the moon, and the highlands, which are full of craters, is incompletely known. The surface mapping that a colony could carry out would be a great improvement over our present knowledge.

Perhaps the most unexpected advantage of a lunar laboratory may be economic. Having spoken of $10 billion in expense, that may seem improbable. However, remarkable possibilities in pure research as well as economies in industrial applications and defense make the enterprise appear promising.

One vital need for defense is to know when any rockets are launched on earth. The present location of the U.S. observation satellites—our eyes in the sky—is in synchronous orbit about seven earth radii above us. The question of how to defend that extraordinarily vital link in our defense system is most difficult. The numerous Soviet satellites on elliptical Molniya orbits appear to be more durable.

One expensive but effective proposal for defending observation satellites, from laser attack and from x-rays produced by a nuclear explosion, is to put a heavy shield—a lot of mass—around them. No matter what the material is, its mass would be useful. However, transportation of such mass is expensive.

That strategic location—the synchronous orbit—is less expensive to reach from the moon than from the earth. As mentioned, the velocity change needed to get into synchronous orbit starting from the moon is one-fourth of the velocity change required for a start from earth. Starting from the moon, the needed fuel weighs less than the payload, whereas starting from earth, approximately ten times as much fuel as payload is needed to get into space. Thus, if energy on the moon becomes available and the project is undertaken from the moon using lunar rocks, the expense of putting protective materials around satellites could be greatly decreased.

Industrial applications also may prove valuable. Some time ago, I proposed that NASA should adopt as its theme song, ''I've got plenty of nothing, and nothing's plenty for me.'' Indeed, an obvious use for the lunar laboratory is connected with nothing—that is, a cheap and excellent vacuum. What sort of vacuum does the moon have? At this time we cannot know accurately, because the astronauts contaminate their immediate surroundings. The lunar colony itself will contaminate the vacuum, but I suspect that it will not have an appreciable effect. The moon itself is emitting gases. Intense ultraviolet solar radiation, micrometeoric impacts, large meteor impacts, and volcanic events all disturb the vacuum.

However, the escape velocity from the moon is so small that the lunar noontime temperature is sufficient for hydrogen to escape. Heavier gases escape more slowly. The atmosphere of the moon—whatever there is of it—rotates with the moon. During the fourteen days of lunar night, it experiences extreme cold and condenses. With the sunrise, the gas again evaporates, rises, and diffuses to the dark side, which acts as a trap. The lunar motion sweeps the atmosphere toward sunrise, and the sun pushes the atmosphere back into the presunrise area. Whatever gas is present is concentrated on one moving longitude—a longitude around the dark edge of sunrise.

Measurements of the moon's atmosphere should be conducted in this area. The pole itself will be quite interesting, because some material might accumulate in its permanently shadowed regions. That is the most likely place for water to be found. The discovery of water deposits would, of course, change many considerations. Then the best way to obtain fuel for propulsion could be through electrolyzing water, using a substantial nuclear reactor or an appropriate solar source of electricity. Thus far, however, the search for water on the moon has proved futile.

Discovering the quality of the vacuum on the moon will have to be done by remote-controled experiments. The search for the best vacuum could be conducted in the lunar night. Should the moon prove to have an excellent vacuum (which is probable), it could lead to one really important application: Surface chemistry could develop from an art into a science.

On earth, breaking a solid into two parts is an irreversible process. The basic reason is that breaking is irregular and distorts the parts so they no longer fit together. Even when the parts do fit—for example, in the case of carefully broken graphite or mica, which come apart in molecularly plane surfaces—the flat surfaces cannot be made to adhere again. The reason is that before the two pieces can be brought into contact, a monomolecular layer of impurities is deposited from the surrounding gas onto the sheared surface; that destroys the possibility for fully effective adhesion.

The moon may possess a vacuum approaching that of interplanetary space—about 1,000 molecules per cubic centimeter. Registering the degree to which mica adheres after being broken and put back together might be an extremely primitive but effective way to detect an excellent vacuum.

Surface chemistry is of obviously great importance in electronics. One of the possibilities that would enable a moon laboratory to pay for itself is that the availability of an excellent vacuum might lift electronics to an entirely new state of perfection. Far more effective chips for computers might first be produced on the moon. The first step toward making such a machine would be to learn more about surfaces.

Some of the early projects for the lunar base would involve making astronomical observations. Mirrors may be composed of small plane elements, adjusted electronically. I suspect that such a telescope placed on the moon would be quite effective. One advantage is that mirrors on the moon can be completely shielded from the earth and the sun. Moonquakes tend generally to be much smaller than earthquakes and would present few difficulties for adjustable mirrors. The main difficulty is that the temperature change between lunar day and night will necessitate careful construction and readjustment of the mirrors unless the mirror is located in a permanently shadowed region near the pole. The main competitor of a lunar mirror is, of course, a space mirror.

The most exciting aspect of lunar and space observations is that they are not limited to visible light, in which the wavelength varies by hardly more than a factor of two. All wavelengths—gamma, x-ray, ultraviolet, infrared, and all radar and radio emissions—are easily observed from the moon. Observations of the sun would also improve, because both the atmospheric interference and the perturbation from the noise originating in our radio emissions would be eliminated.

Astronomy is the most ancient science. It continues to hold public interest. The amount of funding this branch of science has received has been quite limited, yet the recent progress made in astronomy is at least as great as that in any other part of physics. For example, quasars, pulsars, and neutron stars are beautiful and radically new discoveries. The lunar colony, by putting some effort into a study of our own galaxy and of other galaxies, might well improve our knowledge of the origin of the universe. Such basic information is scarce, and the difficulties of finding out more are challenging.

The lunar laboratory may also enrich the study of high-energy physics. At present cosmic rays that have exceedingly high energies are used for exploratory purposes in a limited way. By the time the most energetic particles reach the earth's surface, they are contaminated by interaction with the atmosphere. The best way to begin collecting observations from particles having 1,000 GeV or more might be to dig long collimating holes on the moon. While the frequency of events would be low, the chance of seeing interesting events is a certainty.

One final high-energy physics project deserves consideration: putting an accelerator on the moon. It would consist of separate accelerating units of limited dimensions with long free run for the particles between them. Because of the small size of lunar quakes and the general cleanliness of the environment, such an accelerator is a real possibility. The problem would be to construct the elements that deflect the particles. If that can be done using lunar materials—for instance, to construct the deflecting magnets—a wonderful accelerator could be constructed. It is easy to imagine such an accelerator on the rim of a big crater.

That suggestion allows me to talk about one man who lives on in the minds and hearts of many of us: Enrico Fermi. Fermi, it is said, never showed a slide in his life, and I have tried to emulate that practice. However, the statement about Fermi is slightly exaggerated. He did show one slide, and that was during a talk he gave on accelerators and their probable development. Fermi's slide showed an accelerator encircling the earth. I do not believe that will ever become reality on earth, but our colony on the moon might conceivably complete a smaller version around the earth's satellite.

Forty years ago Robert Oppenheimer talked about having a couple of hundred people working at the weapons laboratory at Los Alamos. By the end of the war, Los Alamos had a population of ten thousand. The lunar base, unfortunately, cannot expand similarly in the near future—the GNP will not allow it. However, I might dare to hope that the lunar laboratory will have twelve people on the five-hundredth anniversary of Columbus's first trip, and one hundred in ten more years. Dr. Hans Mark, the deputy administrator of NASA, has estimated, in analogy with the population of Antarctica, that by the year 2030 the lunar colony may have 10,000 people.

Even with the development of refueling on the moon, the expense of such a colony might approach a trillion dollars a year. Would such an effort be reasonable? In view of the great and varied potential benefits, of which we can now see only the bare beginnings, I would not hesitate to wish it so. We must remember that what we can imagine today may be dwarfed by future realities. There are limits to our imagination. There are hardly any to the developments that the human mind and human activity can accomplish.

24

Machines and Elephants
Never Forget

*   *   *

Author's Note. The following essay was first presented as a speech at the Science and Information Onslaught Symposium, which was held at the Los Alamos National Laboratory on June 4, 1981.

I am indebted to Dorothy Parker for having written, and to Mary Argo for having found, the following poem:

Ballade of Unfortunate Mammals

Love is sharper than stones or sticks;
   Lone as the sea and deeper blue;
Loud in the night as a clock that ticks;
   Longer-lived than the Wandering Jew.
Show me a love was done and through,
   Tell me a kiss escaped its debt!
Son, to your death you will pay your due—
   Women and elephants never forget.

Ever a man, alas, would mix;
   Ever a man, heigh-ho, must woo;
So he's left in the world-old fix;
   Thus is furthered the sale of rue.
Son, your chances are thin and few—
   Won't you ponder before you're set?
Shoot if you must, but hold in view
   Women and elephants never forget.

Down from Caesar past Joynson-Hicks
   Echoes the warning, ever new;
Though they're trained to amusing tricks,
   Gentler, they, than the pigeon's coo,

(*continued*)

189

Careful, son, of the cursed two—
    Either one is a dangerous pet;
Natural history proves it's true—
    Women and elephants never forget.

L'ENVOI
Prince, a precept I'd leave for you,
    Coined in Eden, existing yet:
Skirt the parlor, and shun the zoo—
    Women and elephants never forget.

And what about computing machines? According to the estimates I have, memory capacity in the best machine available today is between 10,000 and 100,000 times smaller than that of a human—male or female. But the cost of memory units is falling rapidly. By the end of this century, and quite possibly sooner, a machine will have as much memory capacity as a human being. What will happen then?

Before I even had a chance to guess, I was scooped by one of my very good friends, Lowell Wood. He asserts that in ten years he will supplement his brain by incorporating a machine in his body. Does Lowell plan to acquire a swelled head? He says not. Where, then, is that extra silicon chip to reside? He replies, "In my belly, of course, where I have lots of room."

When that happens, one consequence is sure: Talk about a "gut feeling" or "bellyaching" will have new connotations. However, before conversation becomes more difficult, a number of other things are apt to happen.

It is hard to imagine a human with a built-in machine in such an intimate association. But the close interaction of a common machine that sits on the table and responds to one's questions is already a reality. The next step may be to ask that machine to identify the various ways in which it could have obtained answers to a question.

That brings up the problem of software, or programming, for the machine. Programming is typically a logical operation. The machine "word" is a collection of information recorded as a series of positives and negatives. Usually the manner of moving from word to word is predetermined by set rules that the programmer has established. The program can be very long and complex; some programs require half a year to write. But once a program is in place, the machine works in a predetermined manner. The future is predetermined by the past, even when the machine is instructed to produce a random number. The fact that the machine will get a random number is predetermined, and even the manner in which the random number is constructed as "random" is precisely prescribed.

Compare that process with human thinking, with human words. The human word has associations. If I have a word in mind, a great number of words loosely associated with it will also be present. Biological thinking, at least as I imagine it, consists of groping around, trying to find associations, and occasionally noticing them.

Thinking is a complicated subject, so my saying something accurate about it is not very likely. Nevertheless, I want to describe what an idea is.

If I start with a word, I find associated words in a diverging array. I may notice that several of those words are associated with words apparently having nothing to do with the word that was my starting point. If I begin to discern order in the words associated with my starting point, the originally divergent process begins to converge. That leads to a pleasurable feeling and to that remarkable statement, "I understand."

A simpler example of the process is connected with humor. Definitions of a joke have often been offered, but they seldom agree. What is a joke? I claim that practically all humor contains an element of the unexpected. The unexpected should not be too painful; if it is, the incident ceases to be a joke. However, if you ask me a question expecting one answer (which you know I know), and I give you an opposite answer, the element of humor is already present.

At this point, I cannot refrain from telling a joke. As behooves a speaker at an information conference, I shall tell a joke about jokes, thereby reaching a slightly higher level of joking. The story is that all jokes can be classified into five categories: first, the jokes the author intends and the audience understands; second, the jokes the author intends and the audience does not understand; third, the jokes the author does not intend but the audience understands; fourth, the jokes that the author does not intend and the audience does not understand; and fifth, all other jokes.

The unexpected is also part of an idea. The man I venerated and loved best as a teacher was Werner Heisenberg. Whenever he heard a new fact, he would ask a question, which I cannot reproduce except in German: "Wo ist der Witz?" Literally translated, "witz" means joke, but Heisenberg meant his question very seriously. His request was for the unexpected element, the essence that summarized the novelty.

When machines with memories comparable to those of humans are available, it might make sense not to program them in a totally logical fashion, but to try to employ the various combinations in which associations can be exploited. A rather rudimentary program might be entered that allows the machine to select and distribute its own activities. By following the manner in which the machine completes its tasks, modifications could be made.

Actually, the S-1 project Lowell Wood is working on at Lawrence Livermore National Laboratory is already attempting to do that in a very primitive manner. The machine is still instructed in a logical form. In fact, the instructions are written in one of the usual programming languages, Pascal. But this composite machine has sixteen arithmetic units, sixteen memory units, and a switchyard connecting them all. The program avoids telling the machine how to distribute the load among all the components. The machine does that by itself. So in a rather loose sense, the machine is beginning to establish its own decisions.

Machines can already play an excellent game of chess, but they are not as good as the best humans. It may be possible to produce a machine that will be able to beat any human, but a more likely unbeatable chess player would be a combination of a good human chess player—Bobby Fischer, for example—and a machine with which he could interact to determine every move. I suspect that such a combination would be able to beat every human and every machine.

I have another proposal, this one involving a formalized kind of joke, a limerick. Limericks have a particular rhythm and rhyme scheme and contain a surprising or unusual idea. I might add that limericks usually have a subject otherwise forbidden, which is part of their humor. That, of course, is still a rather incomplete description of a limerick.

But assume that I instructed a machine to write limericks. I would then conduct an experiment. I would publish a book of limericks, half of them written by a human and half by a machine. If readers could not distinguish between the man-made limericks and the machine-made ones, that would be a fair indication that the computer had a complete list of the rules for limerick writing. Such a list would be a long step toward understanding the nature of a joke.

My definition of a joke is a crude hypothesis that I would like to see perfected. Using machines to test hypotheses about human activities and thought processes could provide valuable insights into psychology. With increased memory and, incidentally, increased speed, such experiments may become a reality. Machine thinking may come to resemble human thought more closely, and we might thereby be able to understand ourselves more fully.

I believe Alan Turing was the first person to suggest that everything a human mind can accomplish, a machine can do faster and more reliably, provided only that the required human thought process is precisely described. That proviso, of course, is fulfilled only in the case of logical human thought—something that occurs with exceptional rarity. The development of machines that could mimic more common varieties of human thought would lift the machine from the category of a servant into an instrument for approaching an understanding of ourselves.

I want to tell yet another story, one that I recall with great pleasure. In the early 1950s, on a pleasant summer day, a group of scientists, including Enrico Fermi and me, were having lunch here at Fuller Lodge.[1] Suddenly Enrico stopped the conversation by asking, "Where are all the people?" In the context of the current discussion, the question made no sense. Yet the seven or eight of us at the table understood him instantly. Fifteen minutes earlier, we had been discussing space travel. Fermi was asking why other living beings had not yet visited us.

Recently I heard a more modern version of the question. If there are other intelligent life forms in the Milky Way system, we should assume that at least some would be ahead of us. Those would have taken robotics to the point where robots were being sent to other planets. Such an ability would lead to an exponential growth of robots all over the Milky Way system. Where are all the robots? Since they are not here, there cannot be intelligent life forms elsewhere in the galaxy.

Speculating that robots would reproduce themselves is only an extension of considering machines that behave in all essential intellectual ways like humans. Yet one could answer the question of where the robots are rather easily. Multiplication on a purposeless basis is wasteful. Perhaps the other intelligent beings have discovered not only robotics but also "family planning" for robots.

This topic is a bit nonsensical, but there is a point to be gained. That humans should be replaceable in their most prized activities— in their intellectual activities—fails to fill most of us with spontaneous enthusiasm. Machines are based on an exceedingly simple foundation—on something hardly more complicated than the mathematical statement that two and two make four. Yet anyone who has looked closely at mathematics knows that playing with numbers is apt to lead to some real surprises. Furthermore, those surprises can make the world appear simpler. That is one of the important points that has been made at this conference.

Are there more surprises in store for us? Will the time come when we know everything? I do not know whether or not there is a single new surprise to be discovered. I certainly cannot prove any answer by logic. But in the history of mental activity, the picture of the world has become more and more integrated, unified, and simple. I have a fervent hope and an illogical conviction that this is a continuing process that will never be complete. Indeed, Gödel's theorem suggests that even something as simple as mathematics is infinite—that it continues to grow new and viable branches.

When I suggest that machines might be used to imitate biological thinking and thereby increase understanding of human thought, I do not mean to claim that the process will ever be complete. I cannot guess whether or not there are parts of ourselves that cannot be approached in this way. It may be that every recognized problem connected with the human thought process can be successfully solved. However, unsuspected new branches may then grow from the fact that the old territory can be more completely explored.

If a future scientist has silicon chips in his belly, he may be stimulated to feats that neither he nor the machine could have accomplished alone. He may be stimulated into ways of behaving that involve thinking on a different level—ways that I can no more imagine than a five-year-old child who has just learned to count can imagine the complex and beautiful theorems that determine which prime numbers can be written as a sum of two squares. My belief is that knowledge and science have no limits and that the interaction of man and machine can play an essential role in the growth of science and the ordering that comes with it.

─── V ───

SCIENCE
AND
RESPONSIBILITY

25

The First Meeting

When I arrived in Leipzig in the fall of 1928, the excitement of studying under Heisenberg was the overriding reality in my life. Heisenberg taught by means of lecture classes and seminars, and the twenty or so of us studying under him in 1929 were an international mixture—a group of Germans, a few Austrians, a Swiss, a Russian, a few Italians, a few Americans, at least one Japanese, and one Hungarian. I learned more and more about the nature of matter, and much of it came through the osmosis of knowledge—the diffusion of facts that spreads among a group of friends. That time marked the beginning of the period of the most complete happiness I have ever experienced.

The youngest of my friends at Leipzig was a German, Carl Friedrich von Weizsäcker, not yet twenty years old. He was close to Heisenberg, extremely quiet, and, in many ways, by far the best of the students. His father had been an admiral in the German Navy, was then a diplomat, and, during the war that was to come, played a role in saving Rome from destruction. Carl Friedrich enjoyed, then as now, walking in the mountains and discussing everything from philosophy to physics, mixing the most serious questions with games or gentle jokes. It was he who helped me to understand a puzzling exchange I had with Niels Bohr.

Heisenberg himself considered Bohr his teacher. After I had spent about a year in Leipzig, Heisenberg advised me to visit Copenhagen. During my first week at Bohr's institute, as a newcomer, I was seated next to Bohr at tea. With all the wisdom of my twenty-one years, I explained some effects of quantum mechanics on educational practices to Bohr, approximately as follows: "Our children will not be taught classical physics at all. Their course will start with the elements of matrix mathematics. Then there will be no contradictions."

I had quite a few explanations and qualifications by which that statement was to be diluted. In the middle of my presentation, I saw that Bohr's eyes were slowly closing. I tried to complete my speech as fast as I could. But Bohr's eyes remained shut after I finished for what seemed like several eternities.

In the end, Bohr, without opening his eyes, gave his reply in a barely audible voice. "We might as well say that we are not drinking tea but dreaming all of this."

Carl Friedrich was there at the tea with me, and on the way home I asked him what Bohr meant. I did not get the answer at that time. It took me a few years to begin to find my orientation in the world of Bohr's paradoxes.

I have come to imagine a similarity between Bohr's comment and one of my favorite passages from Shakespeare:

> Be cheerful sir.
> Our revels now are ended. These our actors
> As I foretold you, were all spirits, and
> Are melted into air, into thin air,
> And, like the baseless fabric of this vision,
> The cloud-clapp'd towers, the gorgeous palaces,
> The solemn temples, the great globe itself,
> Yea, all which it inherit, shall dissolve,
> And, like this insubstantial pageant faded,
> Leave not a rack behind. We are such stuff
> As dreams are made on, and our little life
> Is rounded with a sleep.

Shakespeare and Bohr knew the ambiguous and paradoxical nature of that which we accept as reality. Carl Friedrich has spent his life, as a physicist and philosopher, exploring it. The next chapter comes from a translation of an essay I wrote for the book *Physik, Philosophie und Politik*, which honored Carl Friedrich's seventieth birthday.

26

Understanding

For the past seventy years, Carl Friedrich von Weizsäcker has concerned himself with understanding. In so saying, I mean to suggest that his life span coincides with that activity. While others might say he has concerned himself with physics or philosophy or politics, I see him as having searched for understanding. What does that word mean?

Niels Bohr suggested that the complete meaning of a word could be approached by examining it in more than one language. The process leads to some very curious insights when applied to the word "understanding."

The oral language of the Navaho Indian offers one illustration of what Bohr suggests. The Navaho language was only one of more than five hundred distinct languages used in the New World. The closest approximation to "I understand" seems to be *ak'i'diishtiih,* which is related literally to hearing very well. If a Navaho hears fully, he understands. A description of the meaning is present in the word.

In North America today English, not Navaho, is commonly spoken, and the word "understanding" suggests that you must place yourself under an object or situation or person in order to achieve the process. In other words, modesty is not only useful but also important. The person who thinks he sees everything from a superior perspective probably understands nothing.

In Spanish, *entendimiento,* like the English word "attention," is associated with tension. Nothing can be understood unless it first gains one's attention. Thus the human activity and effort that leads to understanding is emphasized.

The best Latin word for the concept seems to be *comprehensio,* which corresponds to the French *comprehension.* Both words, along with the Russian *Понимание* and the Estonian *Arusaamine,* suggest that what one understands is completely grasped or possessed. The German expression *Begriff* indicates the same connection.

The classical Greek word for understanding, γνώμη, has a straightforward connection to knowledge. It is interesting to note that in modern Greece, the old word is used in a more superficial sense: Today γνώση indicates only an opinion. The passage of time appears to have injected some modesty here.

I find it satisfying that the expression 深知。 in Chinese characters means essentially the same thing as *megértés* in the Hungarian language. The first symbol corresponds to profundity; indeed the strokes remind one of the ocean depths, and the final character corresponds to knowledge, ending with the square, which by itself would represent mouth or word. In Hungarian, *-értés* is associated with knowledge, while the *meg-* emphasizes the thoroughness of that knowledge.

In Bohr's native Danish, understanding is called *forstå*. The German word, *Verstehen,* presents the same curious situation: There is no suggestion of knowledge, but only of a position or a viewpoint. The word is introduced with the syllable *ver-* which has something to do with thoroughness, as in *Vertiefung* (depth), or with resignation, as in *verlieren* (to lose something), *verboten* (to forbid), or *verzichten* (to renounce). One could almost say that the essence of German *verstehen* is to abandon one's, perhaps narrow, viewpoint.

All of this fails to describe what understanding truly is. It only suggests that words are by their nature and essence ambiguous and are associated with all other words. However, understanding has something to do with being able to make connections, as when one translates and hears, with tension, with modesty and the refusal to consider oneself superior, with depth, and words, and knowledge.

On a purely personal note, I can mention the difference between my old friend Carl Friedrich and myself. I appreciate the connotations of the Hungarian word as best conveying the idea of understanding. Thorough knowledge is emphasized. Perhaps my attachment springs from my fear of mistakes. On the other hand, Carl Friedrich appreciates more the breadth and tolerance that is necessary, qualities the German word reflects more clearly.

UNDERSTANDING IN PHYSICS

Breadth of comprehension has proved to be especially important in twentieth-century physics. This has been the period, unforgettable for all of us who participated, in which time and causality came to be understood in a novel and more complete manner. Today I often feel like a stranger among the younger physicists. The idea of complementarity is hardly interesting to most of them. The contradiction that lies at the root of atomic physics and the astonishment about the change in the concept of time seem to have almost completely disappeared behind the mathematical formulation of atomic physics and relativity.

Yet almost all physicists agree that the proper work of physicists is not only to describe the world from a simple viewpoint but also to understand and explain that point of view as a necessary form, actually the only form, of reality that is possible. Whether that commonly accepted purpose should be called a wish or a conviction, I cannot say, but it is clearly a natural continuation of a process of generalization begun in the first decades of the twentieth century.

Carl Friedrich agrees with that goal. He is striving for a fundamental structure of physics based on a simple duality. In logic, it is called yes and no. In atomic physics, the duality corresponds to the twofold possibility of spin direction. It would be wonderful if this great simplification and conclusion of our science would be possible, but I am a bit more skeptical and less hopeful than he.

A new method of approaching the goal is to play with three instead of two possibilities. The colors red, yellow, and blue are used as symbols to describe the strong interaction of the nuclei. Those should behave in a precisely symmetrical way, and the description should progress through the exact science of symmetry that is called group theory. That does not mean, however, that it is possible to replace the dual possibility of spin with the triple possibility of color. Many scientists hope such concepts will lead to the rounding out and completion of physics.

I suspect, however, that we are probably closer to the beginning of physics than to the end. I am reminded of Newton's comment, which I find particularly apt:

> I do not know what I may appear to the world, but to myself I seem to have been only like a boy playing on the seashore and diverting myself in now and then finding a smoother pebble or a prettier shell than ordinary, whilst the great ocean of truth lay all undiscovered before me.[1]

Perhaps the value of understanding lies in the fact, often experienced in physics, that one can use the conclusion of one area of search to find further phenomena and new connections in another—in other words, that knowledge leads to knowledge of another kind.

UNDERSTANDING IN POLITICS

In 1900 few people foresaw the extent of the unfortunate hap-
penings of the twentieth century. Germany certainly has had its full
share of catastrophes. It is not surprising that Carl Friedrich considers the question of how to prevent war the most important. Today
he is fully aware that the tensions are greater than at any other time
in our century. The danger of war seems more horrible to almost
everyone because, undeniably, modern weaponry has a much greater
ability to destroy than ever before.

The pertinent question is: Can understanding help in this situation? Almost everyone is in agreement here too, and it seems almost
by necessity that Carl Friedrich has attempted to use the instrument
of understanding on the biggest political problem of our time. The
problem is not whether understanding is the correct way but rather
how understanding can be useful. One possibility is that if one gains
a thorough understanding of the consequences of a war, one will
strive harder to prevent war. That this approach is being tried is evident. That it will not be sufficient seems highly probable to me.

The required understanding lies less in knowing how to prevent
war and more in learning how to preserve peace. Those two activities
are clearly connected by having the same goal, but the methods they
employ are quite different.

Perhaps it would have been impossible to create lasting peace in
Western Europe without World War II. But the immediate starting
point in the settlement following that war was an agreement regarding the production of coal and iron. That fact may serve as a humble
example of the importance of cooperation.

Two approaches to the problem are necessary, and both must be
followed in order to bring peace. The first activity is to foster cooperative strength among those countries that value peace and stability more than any other goal. The process of disarmament, which
enjoys the greatest popularity in peaceful Western circles, runs contrary to this activity. One must ask: Is it indispensable to continue
in this way?

The second approach is complementary to the first. It is necessary to gain an understanding of the nations and viewpoints that are
basically different from the liberal West. The problem is all the more
difficult because not only one other viewpoint but many other viewpoints are operating among the countless peoples and cultures of our
earth, and several are incompatible with one another.

The viewpoints of the Soviet Union and China, for example, differ both from the West and from each other. They differ not only on such questions as liberalism and freedom of speech, but also in their histories, for China deliberately turned away from technology more than five hundred years ago. The conflict between China and the Soviet Union, which appears advantageous today, may in the long run pose a threat to peace. Furthermore, there is probably no greater contrast than between China, whose heritage of traditions began millennia ago, and the nations of Africa, whose history before 1600 is difficult to trace. Yet in the search for world peace, understanding is equally essential in all cases.

One must look at the garden of humanity in its many details before one can come closer to peace. It is noteworthy and necessary that scientists like Carl Friedrich von Weizsäcker, who in their youth turned to the difficult questions of nature, have to deal in their mature age with understanding human questions. The difference in the application of understanding to the two areas is at least as great as the differences among the descriptions of understanding in the languages of the world.

Yet without understanding, it hardly seems possible to oppose communism or to avoid war. It is not certain that understanding will prove sufficient, but it is certain that without understanding, the situation is hopeless.

UNDERSTANDING IN PHILOSOPHY

In philosophy, there is no understanding; there is only the love of understanding. Philosophers never agree, nor are they expected to agree. Philosophy is not science; rather, it is the origin of science. If a branch of philosophy brings about concrete results, that branch has transformed itself into a science. Understanding is the goal, but it is also the end of philosophy. The same relationship exists between being in love and being married. The origin of the word *philosophy* is not knowledge, but love of knowledge. When a philosophical study becomes serious about knowledge, philosophy is converted to a specialty.

Those remarks are not meant as a value judgment. Philosophy is just as important as science, but science is more immediately useful, and it can be one of the greatest sources of intellectual pleasures.

The word *understanding* has a completely different meaning in pure knowledge.

Carl Friedrich is a physicist, a politician, and a philosopher. I am one of the many who have talked with him about questions in all three areas, and that with the greatest pleasure. My assertion here is almost absurd, because although I am a physicist and have had many political experiences, I do not know very much about philosophy. Fortunately, the nature of philosophy is such that the philosophers themselves do not know much about philosophy. I am grateful to Carl Friedrich for many things, but especially for saving me from positivism. Positivism, with the help of oversimplification, is one of the many ways by which philosophical questions are completely avoided.

Some of the greatest philosophers, including Pythagoras, Plato, and Aristotle, did not confine themselves to philosophy. Those who love knowledge can hardly avoid applying knowledge. Those philosophers' lives show that philosophy is closely associated with science and politics. That association strengthened their philosophies. However, the final result always consists of fragments; understanding will always remain incomplete.

Bohr often mentioned the complementarity of clarity and truth. In a similar way, fruitful ideas and complete thought processes are complementary. Complementarity managed to bring atomic theory to a clear conclusion, but at the same time complementarity demonstrated the fact that one has to use contradictions and paradoxes in order to create essential new thoughts. Complementarity presents an opportunity which remains in the realm of philosophy, because it will always lead to new beginnings and to unusual ways to expand and deepen understanding.

27

An Incident at an Academy

I have found the occupation of being a teacher thoroughly addictive. I began teaching at London City College in 1934, transfered to George Washington University in 1935, gave it up for work on the wartime Manhattan project in 1941, returned to it in 1946, and left it again in 1949. During the 1960s I returned to a little research and more professoring at the University of California, Berkeley. One class I taught was aimed at fighting technological illiteracy.

I am particularly proud of a cartoon published during this period in the student newspaper, the *Daily Californian*. The first panel showed students protesting my political ideas; in the second, the same students were lining up to take my lectures.

I continue to try to fit every invitation to lecture on science into my chronically overly busy life. Children, in particular, are always surprising. A few years ago I lectured at the Hebrew Academy in San Francisco on a fairly regular basis. Many of the children in attendance are refugees from the Soviet Union. One afternoon, when no particular topic was planned, I asked the children what they would like to discuss. A young girl raised her hand and asked: "What can you tell us about UFOs?" I replied that all the sightings probably had some explanation in natural terrestrial phenomena. She remained unsatisfied and rephrased her question—several times. Finally, I asked her why she was so persistent. "Because," she said, "in Russia, they tell us UFOs don't exist, so, of course, they do, and now that I am in the United States, I want to learn about them."

The next chapter is based on a lecture I gave at a conference on education held at Yale University in 1962. It contains a capsule outline of the course that I most enjoyed teaching. In that course, I did not discuss UFOs.

28

Science Appreciation

Our lives, our hopes, our fears, our expectations are quite different today from those our ancestors experienced two hundred years ago, when the process called the Industrial Revolution began. That ever accelerating change, almost equivalent to a mutation, was brought about by science.

Yet the great majority of people today—not just those who have relatively poor educations but even intellectuals—know next to nothing about science. My friends in fields other than science will discuss all kinds of topics, but if I mention an old scientific subject, like the theory of relativity, I see their ear flaps coming down. They know that I am about to be presumptious and talk about things they cannot possibly understand. It is almost as though we spoke different languages, as though we were members of a different civilization.

That is frightening, because we live in a democracy. If we lived in a totalitarian system, the situation would be relatively easy. Only a handful of people at the top would need to be informed in order for the decisions to be based on knowledge. I am dedicated to the idea that the majority of the people should make the decisions. But if people do not understand the issues, they are not likely to make the right decisions.

One familiar example of this problem involves the question of fallout from weapons testing. In 1956 a candidate running for President was asked on national television about fallout. He started his answer with the statement: "I know nothing about nuclear physics, but . . ." and then he gave his opinion, whatever it was. If that man had never read Shakespeare and could have dragged that fact into his answer, would he have wanted to boast of his ignorance? I think he would have known that quite a few people in the audience would not want to vote for an illiterate person. But saying "I know nothing about nuclear physics" is good form. That kind of ignorance does not lose any votes, even if the candidate then goes on to talk about a question connected with nuclear physics. I mention the question of fallout to illustrate that a firm popular opinion can be formed on a completely irrelevant basis. And popular opinion has a powerful influence on the government of our country.

Reasonable decisions concerning our welfare, even our survival, depend on experiences in the classroom. But we cannot convince our children they must learn something we ourselves find uninteresting. What can be done? I really do not know. I am sure the necessary changes cannot be accomplished by a trick or a more clever method of teaching.

What I hope might help is best introduced by saying what I believe about the idea of talent. I do not know what talent is. I am not even sure that such a thing as talent exists, although its effects are obvious—some people are "talented" in music, or mathematics, or experimental science, or languages. The only objective sign of talent that I know is a passionate interest in a subject. The person talented in mathematics is not better at arithmetical problems than someone else. The person talented in mathematics is one who, if he gets hold of a mathematical puzzle, cannot get loose from it; he must solve it or remain unhappy. A person who can solve mathematical problems at lightning speed appears to be a genius. Yet I suspect the explanation of such an impressive performance is the same as when a person understands a joke immediately. The chances are he does so because he has heard that joke or some variation of it before.

Science consists of puzzles. In fact, I sometimes believe that science is not much more than a pyramid of puzzles, each layer standing on the shoulders of the previous layer. Interest in problems is something that, in most people, needs some stimulation in order to flourish. I contend that the first and the greatest duty of a teacher is not to instruct but to discover the interest of the student, and then to feed and stimulate that interest.

We are social animals, and the best thing a teacher can do is to demonstrate to the student that the teacher himself is interested in the topic, that he himself has fun with it. We are closely related to monkeys, and to me that is most obvious when I talk to a roomful of the youngest of our species. Nothing is better than to show them something that I like to do. Since they are really simians, they like to come and play the same game.

The children who are vitally interested in science form a minority. Any special occupation must be a minority. We must convey to that minority the feeling that the best among the mathematicians or the chemists or the physicists are no poorer than the best of the football players. That seems a modest requirement, but I think it's sufficient.

The mass of students will not and should not become chemists or mathematicians or physicists, and in a democracy they too must be well informed. Scientists are human and are therefore most apt to do their best work when they feel it is understood by their society. In the age when the best music was created, music was appreciated by a great many people. In the age when the best paintings were produced, painting was admired by everyone who was anyone. That is an inescapable interaction.

Today, the man in the street says to the scientist: Go ahead and play, but please leave me alone. The result is that the scientists go off in a corner and talk polysyllables to each other until no one understands anything. That is the process by which highbrows are bred. If an integrating element of common interest is not present in a society, then the intellectual branches fragment and starve.

Science, like music or art, is not something that can or should be practiced by everybody. But we want all children to be able to enjoy music, to be able to tell good music from poor music, so we teach them to appreciate music in a discriminating manner. That should be the aim in science education for the nonscientist.

Continuing this comparison, how is music appreciation taught? Somebody might suggest that in order to appreciate music, one at least must play a simple exercise on some instrument, for example, the piano. Music is best when played accurately, and playing the C major scale is a very easy piano exercise. Would anyone suggest that a student should be taught music appreciation by practicing the C major scale until he could play it exceedingly fast without fault? That is how science is taught in our high schools today.

Finding fault is easy; proposing solutions is harder. A great Hungarian humorist, Frigyes Karinthi, wrote a book, *That's How You Write,* in which he parodies his colleagues. He begins with a story about a sergeant inspecting his soldiers at target practice. The sergeant watches the shooting with obvious disapproval, takes the rifle from the hands of the first soldier, shoots at the target himself—and misses by a mile. He shouts, "That's how *you* shoot." On the seventh try he hits the bull's-eye, and then says with a smile, "That's how *I* shoot." Actually, the only way to proceed is by trial and error. I can only tell you how I am trying to shoot. How far off target I am, I don't know.

This year at the University of California, Berkeley, I have begun a course called Physics 10, designed for freshmen who do not want to study any physical science. It has no prerequisites and currently has a large enrollment. The word "teach" does not quite describe what I am doing. I am trying to tell the students a lot, including quite a bit of material that is considered difficult. But I ask only the simplest questions. For example, in the beginning of the course, after offering a little background in geometry, I ask: "If you have a right-angled triangle, which of these two statements is correct: The sum of the two shorter sides is equal to the longer side, or The sum of the squares of the two shorter sides is equal to the square of the longer side?" As far as I know, no one has missed the answer.

I never ask questions about the difficult subjects. The difficult things are there only to expose the students to something that is on the horizon of their comprehension or a little beyond it. In music, instead of only practicing scales, students should listen to the Ninth Symphony, even though they cannot be expected to understand it (whatever that word "understand" means either in music or in science).

However, I have to grade the students in some manner, and so I selected two dozen books. The students are to choose and read some of them. When they have read one, they report to a teaching assistant and convince him that they have read the book, and if possible, that they have understood it. Maybe they will get into the habit and continue their reading after the course is over. I have great hopes and some fears.[1]

After some exceedingly simple statements in geometry, my lecture continues with the geometry of events. In that connection I tell them about Einstein's special relativity. I explain that if two events appear simultaneous to one observer they need not appear so to other observers, and that light appears to move with the same velocity no matter how fast one runs after it.

My son, who is still in high school, obtained permission to audit the course. He is there every Monday, Wednesday, and Friday at eight o'clock playing the exceedingly useful role of stool pigeon. When I made some statements about relativity, he turned to his neighbor and said, "Don't you think that's nuts?" His neighbor said, "I think so, but I don't believe he is to be blamed for it." I can only say that I wish I were.

Then I go on to discuss statics, including the story about Archimedes jumping out of his bath and crying "Eureka" when he discovered the law of buoyancy. I also describe the concept of vectors, work, and torque, and what is meant by a scalar product and a vector product. Probably half of the class understands those concepts. The other half at least carries away the idea that mathematics is concerned with more than what is obvious or completely incomprehensible.

Classical mechanics is the next topic, beginning with the story of how Aristarchus discovered that the earth is moving, was ignored by his contemporaries, and had his theory revived in the Renaissance. Then I tell them about the simple and unsophisticated way in which Newton discovered and used the methods of differentiation. I don't try to be precise. For practical purposes, it is enough to say that velocity is a short distance divided by a little time.

Having finished with Mr. Newton, I try to explain that gravitation is the effect of the curvature of the four-dimensional space-time. I don't say it quite that way. I discuss the idea for an hour. I fear that out of the eight hundred students, about eight have an idea what I am talking about. I hope that this is the worst I do. At least the class has not resigned in a body.

Applying the principles of mechanics to the structure of matter forms the next section of the course. I begin with the statement that there are atoms and molecules, that crystals are put together in a regular way, and liquids are put together like a suitcase packed in a hurry. I then talk about the kinetic theory of gases and the statistical treatment of mechanics. I do not discuss the empirical laws of heat. I think that might be boring. I simply tell them what we believe happens in microscopic mechanics and explain that this corresponds to heat.

Then I go back to a portion of Newtonian mechanics. Starting with the motion of a pendulum, I talk about wave motion and interference. Once that is understood, I can introduce electricity, magnetism, and electromagnetic waves.

Finally, I come to atomic and nuclear structure. Most particularly, I try to explain quantum mechanics and the strange fact that when we get to the smallest things we can no longer say whether they are particles or waves. It is important for our young people to know that the laws of cause and effect really hold only for large aggregates of material, for machines, or bodies like planets or stars that move in well-defined orbits. In the microscopic world, behavior is as capricious as that of humans. The behavior of a single atom cannot be predicted. Yet the average behavior of many atoms obeys strict statistical laws.

The mathematical concepts needed to talk about the structure of the atom are not as difficult as the mathematical methods of relativity and the curvature of space. Nevertheless, the results of quantum mechanics are as surprising as anything we have ever seen. I think everyone should be exposed to those surprises. Just as a man in the sixteenth century had to know that the earth is round, so a man of the twentieth century must know that the space-time continuum is not the thing that most of us had imagined and that the atom is not the little machine that materialistic philosophers dreamed about.

I don't know how much the students get out of the course, but I do know that if I taught the course to kids in high school, I would have an easier job. The greatest difficulty is not learning the new concepts but unlearning old, wrong, deeply ingrained ideas. The earlier a person hears about new concepts, the more easily he understands them.[2]

A story holds that when Carl Friedrich Gauss was about eight years old, his teacher punished his class by giving them the assignment of adding the numbers from one to one hundred. Gauss gazed at the ceiling for a few minutes, wrote down a number on his slate, and laid it on the teacher's desk, face down. The teacher looked at him and thought, "I'll get you; you didn't even try." The rest of the class worked hard. In the end, only Gauss had the correct answer.

How he did it can be understood by any child. Gauss added up one and a hundred, which is a hundred and one; and two and ninety-nine, which is a hundred and one; and three and ninety-eight, which is a hundred and one, and he noticed that he could do the same thing up to fifty and fifty-one, which is also one hundred and one. Therefore, the answer must be fifty times a hundred and one, which is 5,050. That is the correct answer.

His teacher, instead of scolding him, took him in hand and started to teach him some real mathematics—at that early age. By the time he was eighteen, Carl Friedrich Gauss had discovered non-Euclidean geometry. When our students are eighteen, they don't know even Euclidean geometry properly, and by the time they learn it, they are probably too old to invent anything else.

We are living in a world in which our way of life is being put to a severe test. I do not think democracy will survive unless we correct some of our behavior, including our habit of neglecting science education. Democracy has been misinterpreted to mean that the schools must not ask for any accomplishment that is too hard for the average student. To say the same thing unmathematically: We have demanded that the average be much lower than the average. We have demonstrated experimentally that this is possible. We shall not survive unless we demand the opposite, that the average be far above the average. I think that this too is possible.

None of us can be excellent in everything, but each of us must be excellent in something. Unless a person can rightly feel that in some respect he is better than his neighbor, that there is one field in which he can make a unique contribution, he is not a happy person, he is not a whole person, and he is not an educated person. I don't care whether it is in sports, in literature, in philosophy, in mathematics, in science, in art, or in music, but it must be in something. The rest of us should appreciate him.

Excellence and the appreciation of excellence should be the aims of education.

29

The Labyrinth of Simplicity

In the traditional culture, science and technology are viewed as the sources of complication. Large numbers of educated people, even learned people, feel impotent and despairing in a world that seems both changed and burdened by progress and knowledge. They wish the world were simpler, more as it used to be. For them, simplicity begins with wanting to discard science and technology.

About the time the Middle Ages were expiring, the legend of Dr. Faustus became popular. Faustus, an alchemist, gained superhuman powers by selling his soul to the Devil. A great many people considered the accelerating Industrial Revolution to be the Devil's work. Our contemporaries who want to enlighten us on the evils of progress may not realize that they are intellectual descendants of a medieval point of view.

Science has played a part in creating that malaise, but its role differs from what most people imagine. Science has produced new ideas at an incredible rate; the ideas themselves seem incredible. Accumulation of unassimilated information is part of the problem. It has helped to create an indigestion that has the symptoms of anxiety, despair, and even fear.

There can be no question but that modern research heaps observation upon observation and pours new facts into an ocean already brimful of details that no one can remember. The purpose of science is to find simplicity and coherence in the mass of information. In that sense, science serves as a guide. The details may remain in reference libraries or on the memory discs of electronic computers.

What is science? My definition includes a qualification not commonly found in dictionaries. I would say that *science is a set of consistent statements of general validity that also contain an element of surprise.* Science is almost always surprising; it may even be an affront to conventional perceptions. The element of surprise has given rise to difficulties and excitement among scientists. It also has managed to trouble and confuse others.

Most people don't think of science as surprising, but they generally agree that a scientific statement, when first encountered, is not obvious. It will seem obvious only after it is considered in relation to many other facts and explanations. The feeling that the world is too complicated to understand is not uncommon in human history. Incredible information has always been particularly frightening when it deals with ideas that cannot be demonstrated in a simple and direct manner.

One of the general and basic drives in human beings is the desire to understand. The question *why* is an important sign of human nature in children. It is a question that needs a careful response. In my opinion, the best response is the one that will cause the child to continue the questioning process. For most of man's existence, the response to human curiosity has been the invention of an incredible variety of myths and fables. They are charming, fascinating, and often profound, but they lack consistency and simplicity. Science is a fable that has been made consistent.

To the physicist, Einstein's relativity introduces not complication but simplicity. Einstein proposed a surprising new relation between space and time and a similar relation between matter and energy; finally, a generalization of his ideas explained gravitation. The ideas of Niels Bohr about atoms unified the huge fields of physics and chemistry. Rules have been replaced by understanding. The facts in modern science have become more numerous, the guidelines more simple.

The importance of simplicity is not restricted to science. A similar situation is found in the arts. When Mozart was fourteen years old, he listened to a secret mass in Rome, Allegri's *Miserere*. The composition had been guarded as a mystery; the singers were not allowed to transcribe it on pain of excommunication. Mozart heard it only once. He was then able to reproduce the entire score.

Let no one think that it was exclusively a feat of prodigious memory. The mass was a piece of art and, as such, had threads of simplicity. Structure is the essence of art. The child who was to become one of the world's greatest composers may not have been able to remember the details of that complicated work, but he could identify the threads, remember them, and reinvent the details having listened once with consummate attention. Those threads are not easily discovered in music or in science. Indeed, they usually can be discerned only with training and effort. Yet the underlying simplicity exists and, once found, introduces more powerful relationships.

There are many questions about the world that are unsolved. Not surprisingly, they are the biggest questions of all for most people, including me. What is life? What is consciousness? What are human beings? Those questions may never have answers. Someday they may be understood as having been the wrong questions to pose; they may need to be asked in terms that today would appear peculiar to everyone. I like to say that life is a little matter endowed with enormous purposeful complexity.

My answer is unscientific and has much in common with artless myths. Like them, it postulates what it pretends to explain. In early fantasies about creation, the world was explained in terms of gods modeled after humans or even after the animals the gods were going to create. In my definition of life, a little present-day knowledge is included. I use the word "matter" in good conscience, because it has a scientific meaning. In talking about "complication," I am on less certain ground. The adjective "purposeful"—now that word is truly dangerous. What purpose? Whose purpose? At this point, I am completely beyond my depth. What I want to point out is that to be alive implies complication, which is handed on from generation to generation.

Simple observation, medically elaborate observation, even more, observation in molecular biology show that complication is a fundamental property of life. Simplicity is of interest only as an opposite to unnecessary complication; taken by itself, it is nothing more than monotony. The scientific point of view, in its orientation and consistency, differs from myths, legends, and poetry and a host of magnificent human achievements. Science introduces consistency and simplicity into a world that without them appears confused and random.

30

The Science of Paradoxes

Playing with numbers was one of my earliest entertainments. Instead of going to sleep, I tried to figure out how many seconds there were in an hour, day, and year. That I got different answers on different nights made the game more interesting. All of those activities went on before my first collision with the alphabet.

Within my family, no one shared my interest in mathematics, but my father had an older friend, Leopold Klug, who was a retired professor of mathematics. Professor Klug was the first adult I met who seemed to enjoy what he was doing. He gave me a copy of Euler's *Algebra.* I read that work before I became a teenager. The book had been written two centuries earlier.

My first eighteen years were spent in Budapest, close to the Danube. Toward the end of that period, I became friends with three men already practicing science. We had not attended the same schools, but the community in Budapest was small enough that I had a chance to meet them before I finished high school. Two were "older" people—in their early twenties: Eugene Wigner, who became a great theoretical physicist, and Johnny von Neumann, whose brilliance as a mathematician is internationally acknowledged. The third was almost elderly: Leo Szilard, of lesser fame but great ability, was then in his late twenties.

All three of them studied and worked in the scientific community of Germany, which was soon to be destroyed by Hitler. The only topic of our conversations in those days was physics. I still remember Szilard's words that only two questions were of interest: the theory of atoms and the theory of gravitation. Many physicists today would repeat those words with little modification.

Johnny von Neumann had wanted to become a banker like his father, but his father realized that Johnny inevitably would become a mathematician.[1] My father did not have so great a reason to see a mathematician in me. He insisted on a more practical course. I was sent to study chemistry in Karlsruhe. But I cheated. I added mathematics and had a double major.

Toward the end of the second year of my studies, I heard a brilliant lecturer, Herman Mark, talk about the new science of molecular spectroscopy. He made it clear that new ideas in physics had changed chemistry into an important part of the forefront of physics.[2] About that time, my father visited me, and I got his permission to study whatever I liked. By then the new discoveries in the theory of atoms seemed even more enticing than pure mathematics.

What had happened was more than a contradiction between two facts. It was a collision between two systems of facts, between the sciences of chemistry and physics. Chemistry dealt with well-defined and conclusively demonstrated states of atoms and the groups of atoms called molecules. At the beginning of the twentieth century, chemists had identified the composition and structure of almost a million different substances and had developed the ability to synthesize both naturally occurring compounds and others never seen before. All that work was in accordance with an atomic theory that suggested matter was made up of small, immutable components called atoms, which could combine with other atoms to form a molecule, and thereby take on a different set of characteristics. Furthermore, scientists knew that electrical forces were involved in atoms.

In the first years of the twentieth century the size of the atom was definitely established. Atoms were found to consist of a positively charged nucleus, very much smaller than the atom, with electrons loosely arranged throughout the surrounding volume of the atom. Because the negatively charged electrons were not pulled into the positively charged nucleus, it was assumed that the electrons moved around the nucleus like planets around the sun. Physicists who deal with huge aggregates of molecules and their continuously changing behavior in space and time had determined that moving charged particles radiate energy, like an electric antenna. In fact, according to the well-established laws of classical physics, one can calculate that in much less than one-millionth of a second, the electrons within the atoms should radiate away their energy and fall into the nucleus.

Chemists had found that all atoms and molecules of a given type had identical properties. Yet, if atoms and molecules were composed of nuclei and electrons, why didn't they change their intrinsic structure during any of the enormous number of high-velocity collisions they constantly had with other atoms? The final and most bothersome question involved the statistical theory of heat, again a well-established body of scientific information. According to the theory of heat, all parts of the atoms and molecules should accept some heat energy. However, electrons within the atoms and many atoms within molecules seemed not to do so.

The laws of chemistry and the laws of physics were in clear disagreement in those days. It was Niels Bohr who fully recognized the problem. He postulated in 1912 that atoms exist in definite states. He did not explain his postulate but proposed the means by which one could find one's way among contradictory, paradoxical facts. In 1925, Heisenberg found a consistent mathematical form (called quantum mechanics) in which Bohr's ideas could be completed. His proposal developed into a self-consistent theory of the atom and the structure of matter. In 1928, I went to study at the University of Leipzig under Heisenberg, who was only seven years older than I.

Life was full of challenge. The surprises of science are not easy to accept, particularly when one devotes full attention to them. As a high school student, I had written a statement in my Hungarian diary that reflected despair over my confusion:

> What is called understanding is often no more than a state where one has become familiar with what one does not understand.

In the course of time I came to realize that that view is one not of despair but of hope. It is possible to get used to a set of facts by thoroughly exploring their interrelationships. Those strange facts can become so familiar that one no longer feels lost among them. When one can use them to predict, then one can say, "I understand." Understanding is the greatest intellectual comfort.

The process of acquiring understanding is seldom easy. Heisenberg saw to it that his students got to know other famous scientists. So it was that on one occasion his whole seminar, of which I was a part, was taken to Berlin to hear Einstein describe his latest ideas about a unified field theory, which was supposed to explain everything. After his first few introductory remarks, I was completely confused. I couldn't follow a single thought. When the seminar was over, some of us went to the famous Berlin Zoo.

Eugene Wigner was then teaching in Berlin. He had come to hear Einstein speak and joined us on our excursion. It was a beautiful day, and I was conscious of the sunshine. But I was depressed and let my nose hang in a creditable imitation of an elephant. Eugene, being an exceptionally kind man, asked me what was wrong. My answer was to the point: "I am so stupid." Had he tried to contradict me, it would not have helped. But, instead, he told me the truth. "Yes, yes," Eugene said, "stupidity is a general human property."

Whose troubles can stand in the face of such a revelation? That general human property has been demonstrated many times in the Teller family, but somehow calling it by its Wignerian name has eased its woes. Even Einstein had his problems. The ideas he presented that day were wrong. Although he worked on his unified field theory for forty years, he never was able to perfect it. While both atomic physics and Einstein's general relativity suggested that physics was likely to end in recognition of some ultimate truth, the expectation in both cases was disappointed. (Those expectations are as alive as ever in the contemporary physics community.)

However, the atomic revolution introduced a new reality in science. An essential part of the theory of relativity is the importance of the perspective of the observer. That idea is harmonious with an individualistic philosophy and has many parallels in everyday life. In 1900 the German theoretical physicist Max Planck introduced the concept that light is emitted in quanta, that is, in well-defined quantities. Einstein went further and said that light behaved as an assembly of finite particles. Actually, the description of light as a particle or a wave is necessarily incomplete, just as man cannot be described completely either as an assembly of matter or as a pure spirit. That does violate our intuition, but nowhere does it violate logic. The problem in understanding is that most of us find it hard to keep our most elementary intuitions separate from logic.

The son of a French duke, Louis DeBroglie, hypothesized that the wave-particle duality seen in the nature of light was also a characteristic of the electrons found in atoms and molecules. His work was not recognized by his colleagues, but it got Einstein excited. He sent de Broglie's Ph.D. dissertation to some of his friends, including Erwin Schrödinger. Schrödinger found a connection between de Broglie's waves and Bohr's paradoxes, and pointed out that if we consider a circle with waves moving along the circle, the waves should close or wind up on themselves in one circuit. Bohr had postulated earlier that, as is shown by chemical evidence, atoms exist in definite states. However, while the states of matter—solid, liquid, and gas—

can be easily observed, the states of atoms are determined by the various wavelengths of the electrons that will fit into the dimensions of the atom.

Looking at the wave interpretation of electrons, one is confronted with the dilemma that physicists faced in considering light waves. Waves usually bring to mind the vibration of a medium. However, the electron-wave moves in no such medium, and the visualization becomes absolutely impossible when many electrons and additional particles are considered. In this case, the wave motion is mathematically described in multidimensional space. According to Schrödinger's picture, the electron is not a point particle but is smeared out to the extent that it is easier to think of as a cloud surrounding the nucleus of the atom. The cloud has a density, which actually is the probability of finding the electron at a particular point in space.

The probabilities now being considered are not being used in the sense of the probabilities associated with life insurance or the statistical nature of heat. These probabilities are now expressing fundamental uncertainties that can never be removed. Neither more detailed calculations nor more careful measurements will ever give a more certain knowledge. That is something fundamentally different from the probabilities used in the classical world of statistics. The mathematical formalism that enables one to calculate the probabilities does not represent any other physical reality.

It was hard for scientists to accept that situation as the final answer. Einstein, to the end of his life, could not accept the conclusions. He proposed that space and time are linked, that both depend on the observer, and that they are warped by gravitation. But he could not agree that causality should be replaced by probability. He could imagine that God governs the world by any set of rules, but not that He is playing dice.

Niels Bohr and Heisenberg elucidated the meaning of statistics as it applied to atomic theory. The measurement process that determines the configuration of a group of electrons proceeds independently of strict causality. In general, there is no certainty as to which of the possible configurations will be found when one actually measures. Only a probability can be calculated. That limitation is inherent in nature.

The measurement process itself is not governed by the cause-effect relationship. According to the accepted former way of thinking, one could start from a detailed description of a situation in one instant in time, and then calculate, using the rules of causation, the situation at the next instant, when the time difference between the two instances is very small. Thus the universe appeared to be mechanical and predetermined. All that is replaced in the measuring process by a choice, performed by the measurement. Repeated measurements on similar systems can only check that the probabilities were correctly calculated. The measurement cannot be analyzed in time. Neither can one answer, for instance, the question of how an electron is found at a definite spot when, prior to measurement, the probability of finding the electron was distributed over an extended region. We ourselves are not so far removed from the electron. A single, unpredictable quantum could trigger the action of an unpredictable human.

The Bohr–Heisenberg discussions answered the question of whether some other concept than particle or wave could clarify the problem. The answer was No! Bohr never tired of emphasizing that the foundation of science is a straightforward direct observation. Otherwise there would be no way of saying what one was discussing. It is not possible to talk about measurement unless it is described in classical terms. Yet those measurements will lead to uncertain results. Direct contradiction is avoided by noticing that one is trying to make very fine measurements where the measuring instrument interferes with the quantity that one is trying to measure. Bohr was fond of explaining this by analogy. The question, are you asleep? can be answered only at the price of waking up. The probing interferes with the state under observation.

All this may look more like psychology than physics. Certainly the situation is in sharper conflict with what is usually considered reality and common sense than the theory of relativity. Einstein's innovation merely redefined the concepts of time and space, but redefined them in unambiguous terms. It is the ambiguity in the dualistic explanation of electrons that is disturbing. According to the older view of science, full reality is inconsistent with ambiguity. That is too narrow a view of reality. The world is full of paradoxes, particularly in human behavior. Love–hate relationships exist. Even if it is conceded that a love–hate relationship can be resolved by reasoning, it should also be conceded that once this relationship is reduced to love or hate, the outcome of actions will be quite different from what would have resulted from the original love–hate relationship. Is psychological ambivalence real? Is the path of an electron less so?

The Heisenberg principle made the intrinsic uncertainty concrete. Basically, it says that one cannot simultaneously know the position and the momentum of an atomic particle. Knowing the position exactly means that the wave function is sharply peaked around one point in space. Such a wave function then predicts that the electron may have almost any velocity. Similarly, rather precise knowledge of the velocity results in a wave function that predicts that the electron may be found almost anywhere with equal likelihood. In Heisenberg's words, causality is not really violated but causality simply cannot be applied. Causality applies in classical physics only if both the position and the velocity of the particle are known in the same instant. Because that knowledge is limited by nature, the predictive value of causality is correspondingly limited.

Bohr emphasized the idea of complementarity—that one cannot get an objective and complete understanding in many situations unless one starts from two (or more) approaches that appear to be mutually exclusive. He believed the idea should be explained to all eighteen-year-olds. He was convinced that, in all cases that matter, a single approach to understanding is insufficient. Therefore, he felt that the principle of complementarity had wide applicability in daily life as well as in the sciences.

One of the most peculiar and satisfying aspects of atomic theory is that in some ways it brings the methods of such widely differing fields as those considering the behavior of an atom and the behavior of a person closer together. That statement must not be misunderstood. I do not mean to claim that mankind has arrived at an understanding of life or consciousness. All that can be claimed (and Niels Bohr did claim this) is that when and if life or consciousness is understood, the methods that were used to understand the atom will be, in some new manner and on a different plane, needed again in biology.

The collision between physics and chemistry was resolved by discarding the picture of a universe operating like a machine under laws of cause and effect. Despite the initial complexities of quantum mechanics, the unification of chemistry and physics represents a great simplification. The surprising ideas of the new science have led to additional surprising applications in harnessing nuclear energy. The wonder and respect I felt for these ideas as a student has not diminished with the passage of time. I suspect that the methods required to arrive at the science of atomic physics will also be needed for dealing with the problems that the new technologies raise.

31

Wallenstein: A Parable About Arms

Those who forget history are condemned to repeat it. Those who misunderstand history may suffer a worse punishment.

Two of the wars in the twentieth century were so dreadful that they are referred to as World War I and World War II. The seventeenth century was even more full of misery. Conflict was rampant before 1618, and the Peace of Westphalia in 1648 did not bring an end to violence. But the three intervening decades are known as the Thirty Years' War. Few people in Central Europe who were alive to see its beginning survived to see its end. In Germany, the population shrank to one-third.

Dozens of famous personalities were involved: the Holy Roman Emperor, Ferdinand II, upheld the cause of the Catholic Church. Cardinal Richelieu sent 200,000 men into battle for the unity and power of France. The Warrior King, Gustavus Adolphus, died on the field of Luetzen (near Leipzig) while his enraged Swedes fought on to victory.

Few remember the name of Albrecht von Waldstein. Today he is known to history and literature as Wallenstein. He was the Duke of Friedland, a province whose name means "land of peace." Wallenstein acquired Friedland, a blooming, prosperous region in what is now Czechoslovakia, as a result of Ferdinand's expropriation of the lands of rebellious Protestant lords. He governed well and justly and became the richest man in Bohemia.

Wallenstein's mother tongue was Slavic, but his main purpose was to unite Germany. For that, he waged his campaigns and pursued his negotiations. Yet the German princes disliked the Bohemian upstart. Ferdinand II distrusted his negotiations. When Wallenstein was murdered in the middle of the war, the Emperor rewarded his murderers.

In 1626 Wallenstein raised an army of 30,000 men with the explicit understanding that the soldiers would "eat up" the conquered territory. Thus began the starvation and disease that decimated Central Europe. Within a couple of years, Wallenstein's successes had turned a local war into a world war.

That an army should live off the land it occupied was not Wallenstein's invention, but his spectacular successes extended that desolation to an unprecedented degree. The weapon that Wallenstein forged had a life of its own, which it perpetuated by eating up the lives of two out of every three people in its vicinity. Survivors of such an experience recount its evils, even beyond the seventh generation: "No soldier shall, in time of peace, be quartered in any house, without the consent of the owner, nor in time of war, but in a manner to be prescribed by law."[1]

Wallenstein's own tragic end came as a result of the success of his army. The military victories ultimately led to a reshaping of Europe—the end of Danish and Spanish power, the independence of the Netherlands, the rise of France, the full sovereignty of the member states of the Holy Roman Empire. But like Frankenstein's monster, the army destroyed its creator. The war outlived Wallenstein by fourteen bloody years. It was ended by the exhaustion of Central Europe.

I remember a conversation that took place between my two senior Hungarian friends, Leo Szilard and Eugene Wigner, a few months after the discovery of fission. Szilard said that the discovery of fission in Nazi Germany would necessarily lead to the atomic bomb. While Wigner did not contradict Szilard, he wanted to work on controlled nuclear reactors, not on weapons. When I asked why, he quoted: "He who lives by the sword dies by the sword."

Decades later, Szilard argued for negotiated agreements with the Soviets. Wigner argued for defense. Which one had learned the lesson of history?

Many people believe that if weapons are built, weapons will be used—that weapons are an evil that once in existence takes on a life of its own. Does the story of Wallenstein have anything to do with such a belief? Can contemporary attitudes and motivations be understood independent of past history? Are all arms races alike? Is the contest between capitalism and communism similar to a religious conflict?

An army, once established, must be fed. Technology, once developed, keeps growing. But it has no life of its own. We, the people, are alive. We are responsible. The question raised by technology will not be solved by technology. It must be solved by us. And in order to solve it, we must remember history—not for the sake of its conclusions but for the sake of its ambiguities.

32

Morality and Defense

Morality is the cornerstone of any human society; without it, cooperation between members of the society could not exist. Religion is the chief means by which the moral code is established and perpetuated. All human relationships, including those of war, are subject to moral judgment.

Having stated the obvious, it is easy to approach the absurd. If a moral society renounces particular weapons as immoral precisely because they are most powerful, and at the same time an amoral society deploys and is ready to use those weapons, the presence of morality threatens the continued existence of morality.

The roots of my religious convictions lie in the Judaic tradition. Judaic law never requires what the average human being cannot fulfill. The law must be obeyed, but it makes allowance for human imperfection. For instance, the law does not ask that, when injured, one turn the other cheek. But it does insist that one never, under any circumstances, do more harm than was done—one, not two eyes, for an eye. The civil law of a Christian society and the Judaic law are very similar. Obedience to Christian religious commandments requires a degree of selflessness attained by saints.

On that basis, I would say that the commandments taught by the Christian church tend more toward idealism, while Jewish commandments tend more toward realism. Both perspectives are necessary for the well-being of a moral community. Idealism promotes improvement; realism avoids the disaster that would make improvement impossible. Their extremes—romanticism and formalism—have little to contribute to society.

Throughout this century, the United States has attempted to maintain peace. About twenty years ago, because of the technical gap between novel weapons of attack and old means of defense, deterrence of war came to be based exclusively on the threat of retaliation, a policy described as Mutual Assured Destruction. For a Judeo-Christian society, that policy was morally bankrupt from the beginning.

The only escape from the moral dilemma seemed to be to stay ahead of the Soviets in the arms race and thus never seriously have to consider exercising the policy. But this was not the way history developed. From superiority we went to equivalence, only to find that equivalence could not be verified or even defined. Finally we were faced with the probability that, in the quantity and perhaps even in the quality of arms, the Soviet Union was superior.

A few years ago a major change occurred in the technical situation. Through ingenious inventions and through diligent work (more diligent on the Soviet side), it became probable that an effective defense against a nuclear attack could be constructed. Today, Mutual Assured Destruction is becoming bankrupt on a practical as well as a moral basis. Our retaliatory ability is becoming steadily less reliable.

The potential development of defensive weapons presents a new moral question. In their letter of May 1983, the Catholic bishops of the United States clearly stated that weapons are justified when they defend the innocent. It is a pity the bishops did not mention that President Reagan anticipated them by two months when he asked for an emphasis on defense in March 1983. Having agreed that defense is justified, however, the bishops proceeded to condemn all types of nuclear weapons, no matter how employed. That is not an internally consistent argument.

The bishops' declaration was intended as a contribution to an important moral problem; it turned into a political act. That letter is, in fact, an example of what I would call the simplified romantic approach, hardly more than a crude appeal to widespread popular opinion.

What should be opposed both on moral grounds and by technical means are the methods of mass destruction: fire-bombing, use of nuclear weapons against cities, and biological warfare. Every helpful means is needed for the attempt to destroy weapons attacking the innocent—lasers, rockets, small nuclear weapons, or any other effective device. And every moral nation should be cooperatively engaged in making that attempt. Aggression is wrong, whether carried out by bow and arrow or by the hydrogen bomb. Defense is right, whether it uses a stream of particles or the concentrated energy locked in the atomic nucleus.

An agreement on the morality of self-defense would bring us a step closer to solving the most dreadful moral problem of our age. Without such an agreement we cannot count on the survival of the society that holds moral values so strongly that it calls them human rights.

33

The Role of the Scientist

THE SCIENTIFIC REVOLUTION

Science-and-technology is today a single concept. It was not always so. In the ancient Greek civilization, science was remote and impractical. Technology developed and spread independently of science. The effective linking of science and technology is a thoroughly modern invention.

The oldest technology is astronomy. Indeed, astronomers governed agricultural societies. The similarity of the situation in ancient Babylon and Central America is striking. The people who knew how to predict the proper time for planting became leaders. To call those leaders either priests or scientists is misleading. They were empirical scientists, whose technical tools were calendars and who supported their knowledge with stories that had nothing to do with the astronomy of Kepler and Newton.

Metallurgy is another, almost equally ancient, technology. For many generations, a smith was a highly honorific title, a fact suggested by the common occurrence of this surname in several languages. Like astronomy, the art of the smith was empirical, not based on scientific understanding.

Today, one of the most rapidly advancing technologies is called materials science. Synthetic materials superior in quality to anything in existence are being developed, materials that can be adapted in a great many ways to various uses. The possibility is being discussed that less expensive and more durable cars will be made without metallic parts by the twenty-first century. About four millennia ago, the Hittites introduced iron into their weapons. About a century ago, science understood the nature of steel. In the not very distant future, we may be much less dependent on various metals and other rare materials.

The influence of science on technology is obvious. Another important, rapidly developing technology is called electronics. In 1897, scientists discovered the lightweight and strongly charged electron. Less than two decades later, the electron was being used in radio transmissions. Less than eighty years later, computers, communication, and defense are inconceivable without electronics. Nor is the discovery of electronics the only scientific discovery put to practical use within twenty years. Lord Rutherford succeeded in producing a nuclear reaction in 1923. In 1935, he was still firmly convinced that applications were impossible, but a large-scale nuclear reaction was demonstrated in the first atomic bomb explosion on July 15, 1945. The development of nuclear energy for military and economic uses has been restricted not by technology but by political problems.

In contrast, the industrial revolution was triggered not by science but by technology. In 1765, James Watt built what has been called the first steam engine. It is an interesting fact that a contemporary union, the Brotherhood of Hammersmen, forbade Watt to work. He and his accomplishment were rescued by Adam Smith, who believed that freedom was more valuable than rules. Smith invited Watt to work at the University of Glasgow, a home of science that was safe from the Hammersmen's influence.

But Watt's work was technology rather than science. Indeed, Watt's invention was not the first steam engine but the first efficient steam engine. However, contemporary science had not even defined efficiency. That concept was introduced about forty years later by a French scientist, Sadi Carnot. And while Carnot had the proper concept of efficiency, the concept of energy was not completed until the middle of the nineteenth century, when the conservation of both chemical and heat energy was recognized. Technology came first. Science followed several decades later.

In the late nineteenth century, the field of chemical engineering became the first to use scientific research to introduce new technologies in a systematic manner. One famous early center of such applications was in Germany, where I. G. Farben used the scientific principles of organic chemistry to manufacture artificial dyes. Understanding of the atom further stimulated chemistry, and during the 1920s a real architecture of complicated chemical molecules was developed. That in turn resulted in nylon, the artificial silk.

At present, we are in the middle of a period of even greater acceleration in this scientific revolution. Complex giant molecules essential to the understanding of life processes have received detailed chemical explanations. That includes the science of a class of molecules called DNA. These molecules carry the information that constitutes the differences not only between plants, animals, and humans but also between individuals. Changing the basic inherited properties has had obvious important results in both agriculture and chemistry. What nature performs by evolution in millions of years, science and technology is achieving in decades.

The Industrial Revolution has turned into the Scientific Revolution, and this later development cannot be discussed without emotion. It is obvious that the opportunities surpass our imagination. So do the dangers. It may be most proper to talk about a crisis. The Chinese symbol for crisis is composed of the ideograms for danger and opportunity. In the present situation, both are unavoidable.

THE NOBLE LIE

The rapid advancement of science has made our small globe an extremely dangerous place. And who is to blame? Scientists, of course. They should now find the means to rescue us from our superabundance of power. That is their moral responsibility. Furthermore, science has become too complicated; knowledge is advancing too rapidly. The layman can hardly understand the new problems and certainly cannot solve them. Better leave it to the scientists.

Ideas of that kind are often implied. What is, indeed, the responsibility of a scientist? How are his activities related to the concrete, horrible danger that faces us, the possibility of a third world war? When the scientists were told that they were responsible, many were ready to respond. Whey they were told that they were the only ones clever enough to accept the heavy burden, there was little protest.

In the age of science, should the scientist be king? The question is not new. According to Greek tradition, Socrates discussed it, and scientists in turn have read the words ascribed to him. In Plato's *Republic,* Socrates expounds ideas about how an ideal community should be governed. He proposes that all power should be exercised by a small number of people who have received a most rigorous education. They should own no property, devote their lives to the commonwealth, and live together as in a military camp. Having proved through many years of apprenticeship that they are strong, wise, and unselfish, those guardians, as Plato calls them, should take over the unchecked, unlimited direction of the state. Those highly educated, highly dedicated people are more commonly called *philosopher-kings*. In the modern state, such kings would be picked not from the ranks of philosophers but from the ranks of scientists.

In Plato's dialogue, all types of social organization are discussed, including democracy:

> This constitution may be the most beautiful of all. Embroidered as it is with every kind of character it may be thought as beautiful as a coloured dress, embroidered with every kind of flower. And perhaps, as children and women admire dresses of many colours, so many persons will decide in favor of this commonwealth, as the most beautiful.[1]

That statement is milder than the more familiar description, "Democracy is the worst form of government." But it omits Churchill's qualification, "except all those other forms that have been tried from time to time." Actually, Plato considers absolute rule as leading to modifications, limitations, and liberations, which in turn lead to democracy, which eventually leads to chaos, and then back to absolute rule.

The idea of democracy is, of course, not new. But its persistence, in Switzerland for six centuries and in this great country for two hundred years, is a modern accomplishment. The freedoms established under democracy are in contrast to the practices Plato recommends to the philosopher-kings. He advocates that the philosopher-kings practice censorship and, in some essential cases, hide the truth from the citizens of the republic.

> Do you imagine that a believer in Hades and its terrors will be free from all fear of death, and in the day of battle will prefer it to defeat and slavery?
>   Certainly not. . . .
>   Then we shall expunge the following passage, and with it all that are like it:
>   "And those chambers be seen both by mortal men and immortals, Terrible, dank, and mouldering—even to gods an abhorrence."[2]

The passage to be censored is from Homer. The arts should serve the philosopher-kings by providing a more docile populace. The guardians, be they philosophers or scientists, should in no way be restricted by truth:

> To the rulers of the state . . . belongs of right to use falsehood, to deceive either enemies or their own citizens, for the good of the state: and no one else may meddle with this privilege.[3]

One famous example of that practice is the noble life.

> We shall tell our people, in mythical language: You are doubtless all breth-
> ren, as many as inhabit the city, but the God who created you mixed gold
> in the composition of such of you as are qualified to rule, which gives
> them the highest value; while the auxiliaries he made silver an ingredient,
> assigning iron and copper to the cultivators of the soil and the other work-
> men.[4]

The lie is obvious. The nobility is not.

If, indeed, we delegate authority over essential things to a small body of men, we have betrayed the simplest principles of democracy. Any small governing group, whether it be of scientists or philosophers, kings or generals, communists or fascists, aristocrats or even gangsters, will govern in an autocratic manner. That is a basic fact. Firm opposition to autocracy, to government by the few, is the basis of democracy. In some instances the few have governed in the best interest of all. But on the whole, rule by a minority is very good only for that minority.

Any ideological debate in our time is almost necessarily considered from the standpoint of a confrontation between democracy and communism. From that perspective, Plato's minority government appears to resemble communism more than democracy. If one forgets the historical context in which Plato wrote, one may even wonder whether the Republic is a serious proposal or a satire.

THE SCIENTIST IN A DEMOCRACY

Is there any sense in the argument that the scientist should make the decisions? Is it true that because a scientist conceived of the atom and put it to use, he should now say what to do with it? Do we insist that the men who make the laws, the legislators, apply those laws? Or do we rather separate the powers of Congress from that of the judges? Do we insist that our generals who know most about war make the decisions between peace and war? In a democracy we say that powers should be divided, and the ultimate power must belong to the people.

I deeply believe that not practicing democracy, leaving the decisions to scientists, is dangerous for our country. Scientists need not and should not make the decisions about the uses of science and technology. Scientists have been trained in a peculiar manner. They

are faced with surprising and sharply defined situations; they deal
with puzzles that are like chess problems. With all the elements in
hand, they make a choice. It may be difficult, but it can be done.
And once the solution is found, there is no doubt about it.

Political decisions call for different abilities. They call for an
understanding of a great number of facts. They call for decisions
made on the basis of insufficient evidence. More than anything, politics calls for feeling and for compromise—things that the scientist
in his magnificent but sharply circumscribed field has no occasion
to practice. When he deals with the intricate but consistent web of
nature, he may be a genius. In the ever changing world of human
relations, he is a child.

Scientists have responsibilities that are real and great. The scientist must try to understand nature and to extend man's use of that
understanding. When a scientist has learned what he can learn and
built what he can build, his work is not yet done. He must also explain in clear and simple terms what he has found and what he has
constructed. And there his responsibility as a scientist ends. The decision on how to use the results of science is not his. The right and
the duty to make such decisions belong to the people.

As a citizen, the scientist must contribute to decisions. He must
do so no less and no more than a farmer, an artist, or any other
member of our democracy. But what man of conviction can remember the necessary distinctions in the heat of an argument? In an ideal
sense the scientist's responsibility is limited. By being limited it is
enhanced. What he has to do, no one else can do in his stead. And
the last of his jobs, to explain clearly and objectively his results, may
well turn out to be superhuman. Who can be objective? Who can
separate undeniable facts from implied conclusions? Let us not expect too much from scientific objectivity. Let us be content if the
scientist attempts to be honest. Let us not assume that he is unpre-
judiced. But let us require that he name his prejudices. Free debate
between prejudiced advocates is a tortuous road toward truth. But
it has proved more reliable than any straight doctrine.

The scientist has done his full duty only if he becomes a participant of our vital, multicolored democratic society. Our society is
"embroidered with every kind of flower" and of those flowers, science is one of the most beautiful. Its beauty is indeed due to a con-
straining rule, not to possession of truth, not to authority based on
truth, but rather to the endless search for truth and continued shar-

ing of every new truth and every new problem with his fellow citizens.

In discussing science, scientists, and the paradoxical system called democracy, I have disagreed with positions attributed to scientists and sometimes advocated by them. Nevertheless, there is nothing I value more than science. By that I claim no special preeminence or privileges as a ruler. In human life and particularly in the life of a democracy, many points of excellence are of highest value.

For the mountain climber, the conquest of the summit is the only real passion. For the musician, nothing exists even vaguely comparable to the magnificence of music. The tiller of the soil knows in his bones that everything starts and ends in the good earth. For the physician there can be no more exalted profession than to give health and life. A dedicated politician is deeply aware of the fact that he is carrying a superhuman responsibility on his all-too-human shoulders, and that knowledge gives him courage and strength. In our world there must be many perfections.

The perfection of science has a special role, because our age is the age of science. It is science that changes our lives, habits, hopes, and fears from decade to decade. Science is the motor that propels us with increasing speed into an uncertain future. Still, the scientist should not be in the driver's seat.

Today, scientists are actually in control of one nation. In Taiwan, approximately one-half of the highest government offices are held by experts in the natural sciences. Most of them are American-educated chemists, mathematicians, engineers, or physicists. They have helped to improve the living standard on a small, resource-poor island. In 1950 the Taiwanese were no better off than the people on the subcontinent of India. In 1986 they are as prosperous as Israel.

Who should govern is one of mankind's oldest and least tractable questions. The scientists of Taiwan have made a sizable contribution to the well-being of their nation. However, the role of those scientists does not persuade me that in the age of science the scientist should be king. Taiwan's system of government currently resembles Plato's republic in several respects. However, the Taiwanese situation lacks wide applicability as an example of scientist-kings: No one will claim that those scientists are perfect.

Epilogue: Belief in Action

I like activity. In fact, I believe that there is no other way in which to enjoy life in a consistent manner.

I hate doubt, and yet I am certain that doubt is the only way to approach anything worth believing in.

I believe in good. It is an ephemeral and elusive quality. It is the center of my beliefs, but it cannot be strengthened by talking about it.

I believe in evil. It is the property of all those who are certain of truth. Despair and fanaticism, historically sources of incredible destructiveness, are only differing manifestations of evil.

I believe in excellence. It is a basic need of every human soul. All of us can be excellent, because, fortunately, we are exceedingly diverse in our ambitions and talents.

I consider life a miracle. My desire for logical consistency rebels against miracles. Yet I cannot deny, nor do I dream of denying, the miracle of life.

I know that an understanding of life may require an even greater miracle than the creation of life itself. Yet miracles can be recognized only after they happen. Wishing for them does little good. I am forced to reject the present explanations of life on scientific, religious, or any other grounds. Obviously, I do not claim to understand life. I can only attempt to remain responsive in a flexible manner to its various manifestations.

To be alive implies complication, and the most complex of all systems are those composed of aggregates of human beings. I believe that some laws are necessary for people to live together. Those laws contribute most when they are few and carefully thought out.

The life of the spirit is made possible only by the knowledge of imperfection. To perceive imperfection in others is easy but sometimes mistaken. To recognize imperfection in oneself is obviously difficult, but that ability is the important part of one's relationship to perfection. It is difficult for imperfection to serve as the basis of collaboration between people. Yet, as the story of the democracies show, acknowledged imperfection can be a remarkably solid foundation for human cooperation.

Throughout the years I was growing up, Hungary had a great deal of anti-Semitic feeling, and I am a Jew. Much Jewish philosophy was criticized, and as a high school student I asked my grandfather why Jews taught "an eye for an eye."

My grandfather said, "The law does not command one to respond to an injury. It is better not to seek revenge. But the law says that one must never take more than an eye for an eye or a tooth for a tooth. Anger is a human response, and not everyone can be a saint. The man who has just had his tooth knocked out may be very angry. However, the teaching forbids ever returning more damage than was received. And this everyone can obey." A law that cannot be obeyed by most of us humans is no law.

I also remember an instruction from my father, who was a lawyer. When I was a child, he taught me a Latin sentence that he said was the basis of all good laws and good government: *Audiatur et altera pars,* let the other side be heard too. I believe my father's statement was correct. Concerted human action—which is called politics—is full of immense, far-reaching tragedies that have damaged the lives of everyone on the planet. Most would have been avoided had this instruction been commonly heeded.

I was born in Hungary. I knew when I was young that Hungary would fall victim to either fascism or communism. I did not know that my homeland would fall victim to both.

I have spent the longest part of my life in the United States, the land of imperfection. I immeasurably prefer it to Hungary's neighbor, Russia, the land of age-old belief in perfection. With minor, short-lived exceptions, doubt has never been permissible there. The dictatorial imposition of "truth" has led and continues to lead to incredible cruelties.

In my early youth near the beginning of the century, the following distinctions were used to characterize four nations:

In England, everything is permitted except a few things that are forbidden.

In Prussia, everything is forbidden except a few things that are permitted.

In Austro-Hungary (where I lived), everything that is forbidden is permitted.

In Russia, everything that is not forbidden is obligatory.

During my lifetime, little has changed in the last-named country. I am grateful that I have been able to live where my beliefs and actions could be consonant.

The life of the spirit, although it is less visible than the physical world, is even more actual. Here are some examples.

Three years before I was born, the theory of relativity was published. That set in motion a radical change in the physical sciences. The development of quantum mechanics during my young years changed the scientific description of the world from one of mechanical determinism to one in which even the behavior of atoms could not be completely predicted. Atoms were found to be, in some respects, as unpredictable as we imagine ourselves to be.

That understanding of the nature of the world is in harmony with my beliefs. Paradoxes are very real. A single perspective seldom provides reliable understanding. The most permanently pleasurable experiences of my youth consisted in getting acquainted with those unexpected insights.

Whether my beliefs spring from those ideas or are independent of them, I cannot say. In both the physical and the social world, the perception that complementary, seemingly contradictory visions must be acknowledged in order to approach a real solution seems to be borne out with extraordinary consistency.

For years, I worked on science for the reward of understanding and nurtured and enjoyed the greater reward of friendships that arise from the common pursuit of understanding. But by the end of the 1930s, science could no longer be separated from technology and from all the problems and solutions that flow from technology. Eventually that led to conflict: Is it permissible to pursue understanding beyond the limits within which understanding has only predictable and desirable consequences?

Reaching a decision on that question caused painful, difficult doubts. Later events have continued to reinforce the decision that I reached: The possibilities inherent in the laws of nature must be pursued to their resolution. Self-imposed ignorance is even less likely to have predictable, desirable consequences.

I believe that no limits are set to human knowledge; that knowledge leads to power; that power, when it is equally shared among large numbers of people, can be used to benefit everyone. To believe more seems to me foolish; to believe less seems cowardly.

The one great conflict in my life arose when I was in my early forties. The question was whether a hydrogen bomb should be built. The view I held gained political ascendancy, and I played a significant role in the scientific effort that produced the first thermonuclear reaction on earth.

My action was motivated by a firm belief that we must not recoil from knowledge and its fruits. It was also motivated by my opinion that the Soviet Union would proceed with that development regardless of whether we progressed or held back. My guess was verified as correct by a Soviet success almost simultaneous to ours and by Andrei Sakharov's statement in 1974 that the Soviet project began in the summer of 1948, more than a year before work was authorized in the United States.

In the country of imperfection, I believe, knowledge can be used for positive purposes. I believe that the strength that has flowed from the pursuit of knowledge has helped doubt and imperfection to survive in the face of those so certain of truth that for them no other values can exist.

Because of my beliefs in action, I have obtained what I did not desire and lost one of my great joys. I gained a dubious form of fame. Once, on a television show, seconds before the end of the broadcast, an unexpected question was put to me: "How do you want to be remembered?" My prompt and honest answer was and remains: "I do not want to be remembered." As a result of acting on my beliefs, I lost what I wished to retain: friendly fellowship with many of my fellow scientists. The merits of the friendships I retained with people who do not quarrel over a belief lead me to suspect that I may have gained in friendship more than I lost.

While the price has not been small, I suspect that it has been a lesser cost than having to acknowledge that I lacked sufficient courage to speak my beliefs honestly. With one exception in 1945, I have not denied my beliefs and understandings. I continue to regret that at that time I allowed myself to be dissuaded from supporting and circulating a petition to President Truman. We should have demonstrated the atomic bomb by exploding it harmlessly high above Tokyo before using it against targets on the ground. As it turned out, a demonstration probably would have been sufficient to end the war. Had that occurred, today we would be more reasonable and secure.

I detest pride, but I still tend to be proud of one accomplishment. I helped start a laboratory, which continues to work on the one worthwhile contemporary issue: the prolongation of the insecure freedom that is the hallmark of Western civilization. In the last few years, my young friends the scientists in the Lawrence Livermore National Laboratory have concentrated their work on developing protective defenses against nuclear weapons. I could have no greater satisfaction in my life than that belief and action would meet in this enterprise, so sorely needed to produce increased security and a greater opportunity for peace.

I firmly believe that each of us has a duty to live and eventually has the right to die. I believe that death is an end more perfect than the deepest sleep. That is not in contradiction to Jewish tenets. They do not postulate death as the beginning of perfect happiness. I believe, as I wrote in my diary when I was a sixteen-year-old in Hungary, that without death, life would be unbearable.

I believe that people use logic painfully and sparingly. Logic in humans seems a paradox to me. It is both a perversity and one of the most precious human gifts. It is a perversity because it is rare and painful. It is a gift because of its consequences, for example, the brilliance of mathematics.

Logic is the link between physical and spiritual life, and its use makes both possible. Without logic, all spiritual life would resemble chaos. Nevertheless, in our actions, and even in our thoughts, we are guided much less by logic than by our much more ancient biological processes.

I believe that the preservation of peace and the improvement of the lot of all people require faith in the rationality of human beings. If we have that faith and pursue understanding, we have at least the possibility of success. Total security has never been available to anyone. To expect it is unrealistic; to imagine that it can exist is to invite disaster.

I believe the most important aim for humanity at present is to avoid war, dictatorship, and their awful consequences. Survival is not simply the fact that some people will continue to live on earth. That much I consider a certainty for the foreseeable future. Survival of humanity must also include the survival of human dignity.

I believe that no endeavor that is worthwhile is simple in prospect; if it is right, it will be simple in retrospect. Belief in action will accomplish that which does not yet exist but in the future will be taken for granted.

Notes

A Personal Note About War (pp. xiii–xiv)

1. Erich Kästner, *Herz auf Talle* (Zurich: Atrium Verlag, 1929). Reprinted by permission.

Ein Traum Macht Vorschläge

Ich träume—man kann das ja ruhig gestehen—fast nie.
Ich schlafe lieber, sobald ich liege.
Aber kürzlich hab ich trotzdem geträumt, wissen Sie.
Und zwar vom kommenden Kriege.

Aus den Gräben krochen Millionen Männer hervor
(lauter Freiwillige, wie eine Stimme betonte),
die hoben ihre Gewehre zur Schulter empor
und prüften, wen zu erschiessen sich lohnte.

Sie kamen einander entgegen, fertig zum Schuss und stumm . . .
Doch da schrie eine Stimme, als wäre jemand in Not!
Da drehten die Männer, wie auf Kommando, die Flinten herum
und schossen sich selber tot.

Sie fielen um in endlosen Reihn.
Ich träume doch eigentlich nie . . .
Und wer mag das nur gewesen sein,
der so schrie?

1. Three Controversies (pp. 3–11)

1. Andrei D. Sakharov, *Sakharov Speaks* (New York: Knopf, 1974), p. 30.
2. Phillip M. Stern, with Harold P. Green, *The Oppenheimer Case: Security on Trial* (New York: Harper & Row, 1969), p. 113.
3. Statement by David E. Lilienthal, recorded in U.S. Atomic Energy Commission,

In the Matter of J. Robert Oppenheimer: Transcript of Hearing Before Personnel Security Board, Washington, D.C., April 12, 1954 through May 6, 1954 (Washington, D.C.: GPO, 1954), pp. 400, 403.

4. Ronald Reagan, "Peace and National Security," in *Vital Speeches of the Day* (1983), pp. 389–90.

5. "Appeal to All Scientists of the World," *Pravda,* April 10, 1983.

2. Better a Shield than a Sword (pp. 12–22)

1. *Pravda,* February 11, 1967.

2. Ogarkov, N. V. *Istoriya Uchit Bditelnost* (Moscow: Voyenizdat, approved for publication April 8, 1985), p. 49.

3. Igor Nikolaevich Golovin, *I. V. Kurchatov: A Socialist Realist Biography of the Soviet Nuclear Scientist,* trans. William H. Doughtery (Bloomington, Ind.: Selbstverlag Press, 1968), pp. 39–48.

4. In 1959, research on launch detection (over the horizon—OTH) radar was begun in the Soviet Union. By 1961, research on using space satellites for launch detection was under way. Deployment of long-range launch detection radar began in 1971. Research on an improved type of ABM radar began in 1967, and the first model was deployed in 1975.

5. Benjamin Shriever, retired as a full general, is now active in SDI.

4. The Antiweapons (pp. 26–36)

1. His father, Yakov Zeldovich, worked on the Soviet atomic bomb during World War II and later became world renowned in high energy physics and astronomy. The younger Zeldovich has made a significant contribution to the Soviet equivalent of SDI.

6. The Story of the Atomic Bomb (pp. 45–54)

1. Szilard's logic was sound, but, in the case of beryllium, an experimental error of an earlier decade had led him astray.

2. Computers are very useful if they are understood and used with proper moderation. They are not a substitute for ingenuity. One should not believe that what we accomplish with computers, others cannot accomplish without them.

3. Mary Argo spent the rest of her life in Los Alamos. She raised a family and helped to build a community where we had worked during the wartime emergency to build a new weapon, which kept the peace for a generation. Forty years later, she died of a stroke during another hike in the mountains nearby.

7. Seven Hours of Reminiscences (pp. 55–65)

1. Szilard's letter was dated July 4, 1945, while my reply, dated July 2, was written a number of days after I received his. In addition, Szilard had not bothered to fill in my name in his form letter. The explanation is simply that Szilard was a man of many idiosyncrasies. I know of no one, with the possible exception of his wife, with whom he was on a first-name basis.

2. Fermi had lived many years under fascism, and I suspect this may account for his reticence.

3. The Navy generally opposed the use of the bomb without warning. Strauss, in every way a man who loved his country, was also too honest not to expose all the details of what he considered a tragic error. Some weeks before the Hiroshima bombing, Prince Konoye, who was attempting to negotiate a peace through Moscow, received peace overture instructions that were intercepted and decoded by the Navy. They were practically identical to the terms of surrender achieved at the end of the war. Strauss in his memoirs, *Men and Decisions* (New York: Doubleday, 1962), omits no part of that confusing and for him extremely painful period.

4. Captain (later Rear Admiral) William S. Parsons was the scientific representative from the Navy to the Los Alamos Laboratory, and the party was in celebration of his promotion to commodore.

5. After the hearings Oppie remained for many fruitful years as director of the Institute for Advanced Study in Princeton, N.J. Strauss was the chairman of the board of that institute and had earlier been instrumental in securing the directorship for Oppie.

6. Strauss, *Men and Decisions,* pp. 275–79, 443–45.

7. Chevalier has stated that he told Oppie about a scientist, George C. Eltenton, who was trying to obtain information about the bomb because he believed that Oppie should know this in order to prevent such activities from damaging him or the project. To my knowledge, Oppenheimer never contradicted or validated Chevalier's version.

8. Haakon Chevalier, *The Man Who Would Be God* (New York: Putnam, 1959), and *idem, Oppenheimer: The Story of a Friendship* (New York: Braziller, 1965).

8. The Hydrogen Bomb: The Work of Many People (pp. 66–84)

1. U.S. Atomic Energy Commission, *In The Matter Of J. Robert Oppenheimer: Transcript Of Hearing Before Personnel Security Board, Washington, D.C., April 12, 1954, Through May 6, 1954* (Washington, D.C.: U.S. Government Printing Office, 1954), p. 305.

9. Two Stories (pp. 85–92)

1. Andrei D. Sakharov, *Sakharov Speaks* (New York: Knopf, 1974), p. 30.

2. "Fraglos ist Stalin einer der grössten Verbrecher unseres schwer geprüften Jahrhunderts," Andrei D. Sakharov, *Den Frieden retten!,* trans. Cornelia Gerstenmaier (Stuttgart/Bonn: Burg Verlag, 1983), p. 84.

3. Sakharov, *Sakharov Speaks,* p. 31.

4. *Ibid.* p. 33.

5. Andrei D. Sakharov, *Alarm and Hope* (New York: Knopf, 1978), p. 10.

6. Andrei D. Sakharov, "The Danger of Thermonuclear War: An Open Letter to Dr. Sidney Drell," *Foreign Affairs,* Summer 1983, p. 1010.

7. Sakharov, *Sakharov Speaks,* p. 32.

8. Sakharov, *Alarm and Hope,* p. 5.

10. To Win the Peace (pp. 95–99)

1. Spencer R. Weart and Gertrud Weiss Szilard, eds., *Leo Szilard: His Version of the Facts—Selected Recollections and Correspondence* (Cambridge, Mass.: MIT Press, 1978), p. 184.
2. *Ibid.,* p. 185.
3. Lewis L. Strauss, *Men and Decisions* (New York: Doubleday, 1962), p. 185.

16. The Hazards of Radiation (pp. 124–130)

1. An *r* unit is a roentgen, a conventional measure of radiation that represents the amount of radiation that sets free a given number of electrons in a unit volume of air.
2. Egon Lorenz, Leon Jacobson, *et al.* in *Biological Effects of X- and Gamma Radiation,* ed. Raymond Zirkle (New York: McGraw-Hill, 1953), pp. 24–148. Later, in December 1969, J. F. Spaulding, M. R. Brooks, and G. L. Tietjen reported a similar result (in *Genetics* 63, pp. 897–906). They had exposed many generations of mice to 200 rads (which are in this case approximately equal to 200 *r*) per generation in small doses. There were no effects on viability or fertility, except that the line with ten generations of males exposed to x-rays survived significantly longer than the controls who were not exposed to radiation. Again, this cannot be used to indicate beneficial effects for radiation. All that can be said is that the mice receiving low-dose rate radiation showed no ill effects.
3. W. L. Russell, L. B. Russell, and M. B. Cupp. *Proceedings of the National Academy of Sciences* 45 (Washington, D.C.: National Academy of Sciences, 1959), pp. 18–23.
4. It is popular to oppose exposure to radiation; it is also popular to conserve energy. As Dixy Lee Ray points out in *World Media Report* (Vol. 1, No. 1, Winter 1986), those two notions are in conflict. In order to conserve energy, homes have been made as airtight as possible. That leads to the concentration of radon—a naturally occurring radioactive gas that delivers radiation to the lungs—within well-insulated houses. In August 1986 the Environmental Protection Agency (EPA) prescribed a permissible level of radiation for homes: It is 40 times *higher* than the level allowed for workers in the nuclear industry. In the same report, the EPA estimated that the high limit set for radon is exceeded in 8 million energy-conscious American homes.

17. Widespread Aftereffects of a Large-Scale Nuclear War (pp. 131–142)

1. O. R. Turco, P. B. Toon, T. P. Ackerman, T. P. Pollack, and C. Sagan, *Science* 222 (1983): 1283–1292.
2. Nevil Shute, *On the Beach* (New York: Morrow, 1957).
3. *Long-Term Worldwide Effects of Multiple Nuclear-Weapons Detonations* (Washington, D.C.: National Academy of Sciences, 1975).
4. J. B. Knox, *Global Scale Deposition of Radioactivity from a Large Scale Exchange,* LLNL Report UCRL-89907, 1983.
5. *Ibid.*
6. National Committee on Radiological Protection and Measurement, Report 29

(Washington, D.C.: Government Printing Office, 1962), and *The Medical Effects of Nuclear War* (London: British Medical Association; New York: Wiley, 1983).

7. Knox, *Global Scale Deposition.*

8. *Ibid.*

9. *Long-Term World Wide Effects* and, M. C. MacCracken and J. S. Chang, *A Preliminary Study of the Potential Chemical and Climatic Effects of Atmospheric Nuclear Explosions,* LLNL Report UCRL-51653, 1975.

10. J. S. Chang, W. H. Duewer, and D. J. Wuebbles, *Journal of Geophysical Research* (1979): 84, 1755 and F. M. Luther, *Nuclear War: Short-Term Chemical and Radiative Effects of Stratospheric Injections,* LLNL Report UCRL-89957, 1983.

11. Luther, *Nuclear War.*

12. M. M. Caldwell, *BioScience* 29 (1979): 520.

13. *The Effects of Nuclear War* (Washington, D.C.: Congressional Office of Technical Assessment, 1979).

14. V. C. LaMarche, Jr. and K. K. Hirschboeck, *Nature* 307 (1984): 121.

15. *Long-Term Worldwide Effects.*

16. P. J. *Crutzen,* and J. W. Birks, *Ambio* 11 (1982): 15.

17. Turco *et al.* in *Science* (note 1).

18. C. Sagan, *Foreign Affairs* 62, (Winter 1983): 257–92.

19. *Ibid.*

20. Turco *et al., in Science.*

21. Knox, *Global Scale Deposition.*

22. W. D. Sellers, *Physical Climatology* (Chicago: University of Chicago Press, 1965).

23. W. Seiler, and P. J. Crutzen, *Climatic Change* 2, (1980): 207–47.

24. Turco *et al., in Science.*

25. University of Washington, Seattle, private communication.

26. G. Hanel, *Atmospheric Environment,* 16, (1982): 1703–2706.26.

27. Turco *et al., in Science.*

28. C. Covey, S. H. Schneider, and S. L. Thompson, *Nature* 308 (1984): 21–25.

29. Turco *et al., in Science.*

30. A. A. Broyles, *American Journal of Physics,* 54 (1985): 323–32.

31. C. D. Tangren, C. K. McMahon, P. W. Ryan, and B. T. Zinn in *Southern Forestry Smoke Management Guidebook* (Asheville, N.C.: USDA Forest Services, Southeastern Forest Experimental Station, 1976).

32. C. P. Bankston, R. A. Cassanova, E. A. Powell, and B. T. Zinn, *Review of Smoke Particulate Properties Data for Burning Natural and Synthetic Materials,* NTIS PB 292066, 1978.

33. A. A. Brown, and K. P. Davis, *Forest Fire Control and Use* (New York: McGraw-Hall, 1973), and D. A. Larson and R. D. Small, *Analysis of the Large Urban Fire Environment,* Part 2, NTIS AD A125-400/2, 1982.

34. Turco *et al.,* in *Science.*

35. C. G. Culver, *American Society of Civil Engineering Structural Division Journal* 102 (1976): 2269-84, and L. A. Issen, *Single Family Residential Fire and Live Loads Survey* NTIS PB 81-166647, 1980.

36. Oil Market Report May 2, 1984 (Paris: International Energy Administration, 1984).

37. I thank the scientists in the LLNL Atmospheric Sciences Division—Joseph B. Knox, Michael C. MacCracken, Frederick M. Luther, and Joyce E. Penner—for the research on which my main conclusions are based and for their comments on this paper. Particular thanks are due Professor Arthur A. Broyles, University of Florida. This work was performed under the auspices of the U.S. Department of Energy under contract W-7405-Eng-48. The views and opinions are nevertheless those of the author alone, and the usual U.S. government disclaimer of legal responsibility, explicit or implied, should be understood as accompanying this article.

Author's Postscript

38. L. W. Alvarez, W. Alvarez, F. Asaro, and H. V. Michel, *Science* 208, p. 1095 (1980), and W. Alvarez, F. Asaro, H. B. Michel, and L. W. Alvarez, *Science* 216, p. 886 (1982), and W. Alvarez, L. W. Alvarez, F. Asaro, and H. V. Michel, *Geological Society of America,* Special Paper 90, p. 305 (1982).

39. W. S. Wolbach, R. S. Lewis, and E. Anders, *Science* 230, pp. 167-70 (1985).

20. Chernobyl and the Safety of Nuclear Reactors (pp. 158-168)

1. Pyotr Avanesov, "Accident," *Soviet Life,* No. 9, September 1986, p. 34.

2. This list does not exhaust potential problems. Current reactors operate with neutrons that have been slowed before they cause fission. Reactors can be designed to operate with neutrons that are slowed to only a small extent. The safety of such fast reactors can be ensured, but they introduce a whole new set of problems.

3. In the end, the Hanford reactors were closed down. A new reactor with far less graphite (so the loss of water simply stops the chain reaction) was built.

4. In humans, iodine is deposited in the thyroid gland, cesium in muscle, and strontium in the bones.

5. "Rem" stands for roentgen equivalent in man; it has been adjusted to standardize the biological effects of various types of radiation.

21. The China Syndrome: A Cautionary Tale (pp. 169-174)

1. Much of the detailed information in this chapter is drawn from the magnificent work of Joseph Needham, *Science and Civilization in China* (London and New York: Cambridge University Press, 1971), particularly, in Volume 4, Part III.

24. Machines and Elephants Never Forget (pp. 189-194)

1. Fuller Lodge, inherited from the boys' school that built it, has served as the social center of Los Alamos since its inception.

26. Understanding (pp. 199–204)

1. This quotation is included in Arthur Koestler's delightful book *The Sleepwalkers* (New York: Macmillan, 1959).

28. Science Appreciation (pp. 206–212)

1. The list included twenty-six books, among them Charles Vernon Boys, *Soap Bubbles and the Forces Which Mold Them* (New York: Anchor, 1959); Henri Poincaré, *Science and Method* (New York: Dover, 1952); George Gamow, *Matter, Earth and Sky* (Fort Lee, N.J.: Prentice-Hall, 1958); Donald Hughes, *The Neutron Story* (New York: Anchor, 1959); and Kees Boeke, *Cosmic View* (New York: John Day, 1957), which consists of forty pictures on different scales, from the atomic nucleus to a galaxy.

2. Recently, two decades after the event, a neighbor on an airplane ride introduced himself as a student of the Physics 10 course. It was a happy day for me.

30. The Science of Paradoxes (pp. 216–222)

1. Quite a few years later, Johnny, established as a great mathematician at Princeton, managed to sneak into economics. He wrote the standard text on game theory.

2. Many years later, I was able to return the favor in part by teaching Herman's son, Hans (who was not yet born at the time of the Karlsruhe lectures), some additional quantum mechanics.

31. Wallenstein: A Parable About Arms (pp. 223–225)

1. Third Amendment to the U.S. Constitution, ratified in 1791.

33. The Role of the Scientists (pp. 229–235)

1. *The Republic of Plato,* trans. J. L. Davies, and D. J. Vaughn (London, 1921), p. 557.

2. *Ibid.,* p. 386.

3. *Ibid.,* p. 389.

4. *Ibid.,* p. 415.

Index